Emily Gregory

Elements of plant anatomy

Emily Gregory

Elements of plant anatomy

ISBN/EAN: 9783337278311

Printed in Europe, USA, Canada, Australia, Japan

Cover: Foto ©berggeist007 / pixelio.de

More available books at **www.hansebooks.com**

ELEMENTS

OF

PLANT ANATOMY

BY

EMILY L. GREGORY, Ph.D.

PROFESSOR OF BOTANY IN BARNARD COLLEGE

BOSTON, U.S.A., AND LONDON
PUBLISHED BY GINN & COMPANY
1895

PREFACE.

THIS book contains the substance of the lectures given to the classes in the last half of the second year's course in botany, at Barnard College. The method followed in this course differs somewhat from that generally adopted in either home or foreign colleges. The study of botany, as a science, is comparatively new in this country, and therefore we have the advantage of the experience of Europe, where the science of botany has long held a place equal in rank with that of its related subjects.

In these older institutions the various divisions of morphology and physiology are taught under a single head, General Botany ; the special work coming after this falls naturally into two divisions, systematic and physiological botany. The preparation for these two special departments must naturally differ somewhat in character. For the former, one must have a thorough training in organography; for the latter, an equally thorough one in anatomy. Accordingly the more recent foreign text-books are arranged in two parts, morphology, or organography, including descriptive and developmental botany, and anatomy and physiology.

With us the tendency has been to recognize only the two general divisions, systematic and physiological botany, and to enter upon these departments with very little special training for either. Especially in this true of the study of physiology, a field in which we have made so little progress that very few people understand what is meant by the expression plant physiology.

As a knowledge of the inner structure of plants is necessary to successful work in physiology, it is believed that a brief course in the elements of anatomy should precede the study of physiology. There are few books written in English which are available to the student of plant anatomy. De Bary's "Comparative Anatomy of Phanerogams and Higher Cryptogams" is invaluable as a reference book, but almost useless to a beginner. The design of the present little volume is to furnish a brief outline of the elementary principles of anatomy in a form available to all students of botany who wish to use this science in any direction. For this purpose it is hoped it will meet the wants of several classes: those who expect to follow it with a more extended course in anatomy and physiology; those who will take up the other line of work, descriptive and systematic botany; and finally, students of pharmacy for whom the importance of a practical knowledge of the structure of roots, stems, and other parts of plants, can hardly be over-estimated.

Aside from this, there is only one other feature which serves as an excuse for the appearance of another text-book, that is, the effort to present the subject from the developmental point of view. The experience of several years has demonstrated the superiority of this method over that in common use. It is hoped that a longer experience will serve to improve the present plan, not only by the eradication of its faults, but also by the more complete development of the idea underlying it. It may also be found practicable, at a later date, to combine with the present volume an elementary treatise on plant physiology, and possibly to supplement both by a more advanced work on each subject separately. However this may be, it is quite certain that the measure of our progress in any science may be found in our ability to adapt the thought and experience of other nations to our special needs and resources.

The plates used are taken mostly from the text-books of J. Wiesner and Th. Hartig to both of whom the author hereby expresses most hearty thanks, as well as to the numerous friends who have aided and assisted in the preparation of the work.

EMILY L. GREGORY.

New York, June, 1895.

CONTENTS

PART II.

ANATOMY OF TISSUES.

CHAPTER IV. — TISSUES AND SYSTEMS.

CHAPTER V. — ANATOMY OF THALLOPHYTES.

CHAPTER VI. — ANATOMY OF CORMOPHYTES.

ELEMENTS OF PLANT ANATOMY.

PART I.

ANATOMY OF THE CELL.

CHAPTER I. — THE VEGETABLE CELL IN GENERAL.

1. General Description of the Plant Cell.

A THIN cross-section from the stem or leaf of any plant shows, when magnified, a network of cells not unlike those of the honeycomb. This fact was first discovered in 1667 by Robert Hooke, an Englishman, who happened to take such a section to test the improvements he was making on the microscope. The first real study of cell-structure was made by Malpighi, an Italian, in the year 1671. The section thus examined appears to be divided into small chambers or cavities, separated from each other by a common wall. The single cavity with its enclosing wall, like a room in a house, received the name of cell. The origin of these cells, or elements of plant-structure, was at first supposed to be similar to that of air bubbles in a somewhat viscous liquid; but this supposition was soon found untenable, as in no young growing tissues was there found any indication of the liquid in which the bubbles were supposed to form.

At a much later period it was discovered that the wall, or membrane, which gave the name to the cavity which it surrounds, was really the less important part and that the cell contents were the only necessary element. This is shown by the fact that the wall is a product of the contents, and that at certain periods of the plant's life the cell may exist without it. Discoveries of this kind gave rise to an entirely different conception of the nature of the plant-cell. It is now known that this, in its simplest or least differentiated condition, consists of a small portion of the viscous liquid known as protoplasm, in which, under ordinary magnification, no structures are visible. It is in this general sense that Reinke defines a plant cell as follows: — "An individualized, not farther divisible structure, consisting of or containing protoplasm which either shows life processes or has shown them."

In studying the anatomy of a plant cell it will be necessary to consider one in its ordinary condition of development, that is, as an element of any plant, differentiated sufficiently to perform the ordinary functions of plant cells. Such cells are usually considered as consisting of two parts, wall and contents, or as it is frequently stated, wall and protoplasm, the latter including a nucleus and one or more vacuoles. Before taking up the study of these parts separately, it may be well to examine the cell as a whole, in reference to several features, namely: — size, form, mechanical and physiological principles, and finally to discuss briefly certain theories concerning organized structures in general.

By far the greater number of plant cells are microscopic, but they vary greatly in size. The smallest occur among the organisms known as bacteria. Some of these are spherical in form and measure from seven-tenths to one micro-millimeter in diameter. Among the largest plant cells may be mentioned those forming the internodes of the stem in the group Characeae, where the cell is often several inches long. Pollen tubes, which are also single cells, often attain a length of several inches, and the milk tubes of certain plants are said to reach even a greater length than this.

Cells vary as much in form as in size; those without a membrane incline to the spherical shape, since the protoplasm composing them is in a half-liquid state. Many swarm-spores are pear-shaped, but they generally assume a spherical form on coming to rest. The forms of some naked cells are subject to rapid change, for example the spores of Bangia, a red seaweed.

In all the higher plants, new cells are formed by the growth of walls across the cavities of the old cells. The new walls join the old at certain angles, and when the cells are young, they are inclined to a hexagonal form. As growth continues the form is liable to change in various ways. If the cell should grow equally fast in all its parts, it would tend to retain its original form. This very rarely happens, and even when it does, the shape of such a cell is influenced in a greater or less degree by the manner of growth of those surrounding it, as the growing wall is flexible and its shape easily changed by pressure or traction from without.

The individuality of the cell is shown by the fact that each has its own predetermined manner of development. All young cells of any plant are, at first, nearly similar in form and size, but later on each cell is seen to follow certain laws of growth which are, to a certain extent, independent of all external forces. From these laws, together with various mechanical causes, arises the great variety of form in the cells of ordinary plants. The peculiar forms common to certain unicellular plants illustrate even better than those of higher ones the inherent tendency of cells to grow in a certain manner. Examples of these are the branched mycelium of Mucor, which is one-celled till reproduction takes place; also the many singular shapes assumed by one-celled algae. Of the latter, Caulerpa is perhaps the most wonderful, where stem, leaf, and root of higher plants are simulated by the branching of a single cell.

From the small size of the average cell, two advantages result to the plant. First, strength and solidity; secondly, the greatest possible amount of surface for the transfer of cell contents. The first insures mechanical support; the second is connected with those changes in the chemical nature of the cell contents, by which the life processes of the plant as a whole are carried on. The discussion of such changes belongs to plant physiology rather than to anatomy.

2. Molecular Structure of Organic Substance.

The material of which the plant is composed possesses qualities which are not found in inorganic substances. In order to understand some of the theories respecting the origin, growth, development, and function of the plant cell, it will be necessary to understand some of these qualities, but for a more complete treatment of such questions the student is referred to works on plant physiology.

All substances are said to consist of smallest particles, called atoms, which are indivisible and separated from each other by a small portion of space. The atoms, even of the elements, are supposed to be united in groups, and generally in pairs. When a certain number of atoms of different elements unite together forming a whole, this result is called a chemical compound. The union of these atoms is caused by the attraction which they possess for each other, — the stronger the attraction, the stronger the union. The smallest particle of such a compound, or of an elementary substance which can exist by itself uncombined, is called a molecule. Its nature depends not only upon the atoms composing it but also upon the manner in which they are grouped together, or their positions in respect to each other. A substance may therefore undergo a change of character without change in the number and kind of atoms composing it.

There are some organic substances whose molecules do not unite themselves into groups of any special character. When these are placed in water, solution takes place. The generally accepted explanation of this process is that the molecules of the substance have a stronger attraction for the water molecules than for each other and hence are torn apart and mingled with the water molecules till a homogeneous mixture results.

In the greater number of organic substances, however, the molecules are supposed to unite themselves into groups named micellae, which are considered the units of organized substances. A micella is a group of molecules so closely united that it is claimed by some authorities that water molecules are not able to force their way between them. The micellae themselves, on the other hand, are surrounded by thin films of water. At least, this is their condition when the substance which they compose is in a fresh and active state. If this substance is allowed to become dry by evaporation it shrinks in size; if then again placed in contact with water, the micellae composing it possess so strong an attraction for the water molecules that it sucks up the water: the films surrounding the micellae thus become thicker, and the substance increases in size. This process of taking in water with corresponding increase in size is called imbibition, and the quality belongs only to organized substances.

A process somewhat similar to this, and often mistaken for it, is known as capillarity, which consists of the penetration of water molecules into narrow channels already existing and filled with air. The water simply drives out the air and takes its place, and this process is therefore not accompanied by increase in size. A common unglazed earthen jar placed in contact with water shows this action of capillarity in unorganized substance. The same process may take place in organized matter, as, for example, in a piece of wood containing ducts or tracheae. If this is placed in water, the air may be driven out of the small tubes by the incoming water, which fills the channels before occupied by air without affecting the size of the wood.

It is not known with certainty to what extent the process of evaporation can be carried. It is supposed, however, that under certain conditions organized structures may lose all the water which ordinarily forms the films about the micellae, so that the latter may lie in actual contact; and that when water is again placed in reach of the structure, its molecules have to make an opening for themselves by forcing the micellae apart. Nothing is definitely known concerning the shape of these micellae. There are several reasons for supposing them to be prismatic rather than spherical. That they lie in contact after extreme evaporation is indicated by the non-appearance of air, which if present would cause a difference in the refraction of light. This contact of surfaces could not take place were the micellae spherical. Their behavior under the action of polarized light indicates crystalline surfaces and lines of cleavage. Finally the shrinking and swelling of wood in different proportions, longitudinally, radially, and tangentially, indicate a difference in the three axes of dimension in the micellae.

In order to apply this theory concerning organized structure, the cell must be studied in its various stages of growth beginning with its first formation near the growing tip of the plant. Here the cavity of the cell is seen to be filled with protoplasm which appears to be nearly, if not quite, homogeneous. As the cell becomes older, one or more vacuoles make their appearance. These vacuoles are simply water or sap-filled spaces which arise from the cell taking in more water than can be held in the interstices between the micellae of the protoplasm.

The theory respecting the physical nature of protoplasm is, that in its active state its units of structure are surrounded by films of water, which may be increased to a certain thickness; when this limit is reached no more water can be held, and if the current is continued, the surplus escapes from between the micellae and collects in some part of the cell where it forms a vacuole. This is rendered possible by the nature of the cell wall, which at this stage of growth is not only flexible but elastic. If only one vacuole is formed, it usually occupies the central portion of the cell, and as it increases in size, presses back the protoplasm against the wall so that the protoplasmic contents now assume the form of a layer of greater or less

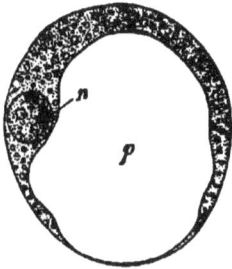

FIG. 1.

Cell with nucleus *n* near the wall. The vacuole *p* contains cell sap by which the protoplasmic body is pressed back against the wall causing turgescence.— *Theodore Hartig.*

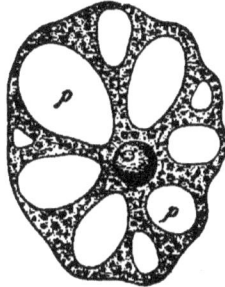

FIG. 2.

Protoplasm containing several vacuoles, nucleus central.— (*T. H.*)

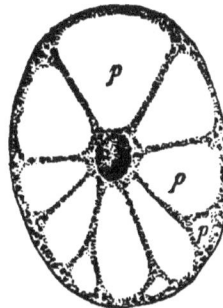

FIG. 3.

Protoplasm in fine strings running through cavities containing cell sap. — (*T. H.*)

thickness, lining the wall and pressing firmly against it. In such cases the nucleus lies imbedded in this layer and is therefore called lateral.[1]

If several vacuoles are formed at the same time, the nucleus remains near the centre of the cell between the vacuoles but always surrounded by protoplasm. When the vacuoles are large the protoplasm between them becomes reduced to mere threads, so that the central portion appears crossed by fine filaments. Outward against the cellulose wall, as well as inward toward the cell-sap, the protoplasm is bounded by a very

[1] The name *primordial utricle* is sometimes applied to this layer of protoplasm, but the term is falling into disuse and the different parts of the protoplasmic mass have received new names corresponding with their nature.

thin membrane, which differs from it in physical structure. This membrane exists also on the outer part of the fine protoplasmic threads, and plays an important part in the processes of plant nutrition. While the discussion of these processes belongs properly to physiology rather than to anatomy, the two subjects are here so closely connected that it is impossible to consider one without including the other.

The plant cell in its ordinary active condition consists of, at least, these four parts: wall, protoplasm, nucleus, and vacuole. According to the explanation given of the organized structure of the wall, it is readily seen that only substances in solution can pass in and out of the cell without injuring it. The substance dissolved may be either organic or inorganic. Very little is definitely known concerning the actual method by which this exchange of material is effected. Experiments show, however, that the protoplasmic membrane is the deciding factor as to what substances may find their way in and out of the cell.

Various substances are found dissolved in the water of the vacuoles. These may be inorganic salts which have been taken into the cell by osmosis, or organic substances formed there by the metabolism of the living cell, or taken in from without as the product of some other organism.

Now if such a cell be placed in contact with pure water, or with a solution whose concentration is less than that of the liquid in the vacuole, the laws of diffusion are such as to cause a flow of water into the cell. In consequence of this incoming water the cell becomes increased in size and the wall stretched and distended. When sufficient water has been taken in to cause the pressure on the stretched wall to be greater than that of the atmosphere without, this condition is known as turgor,[1] and the cell is said to be turgescent.

[1] For a more extended description of turgor, see Pfeffer, "Zur Kenntnis der Plasmahaut und der Vacuolen," page 207. The author in speaking of the

On the other hand, if the cell is placed in contact with a solution of greater density than that of the cell-sap within, more liquid will flow out of the cell than into it, in a given time. If this process continues long enough, the vacuoles will become

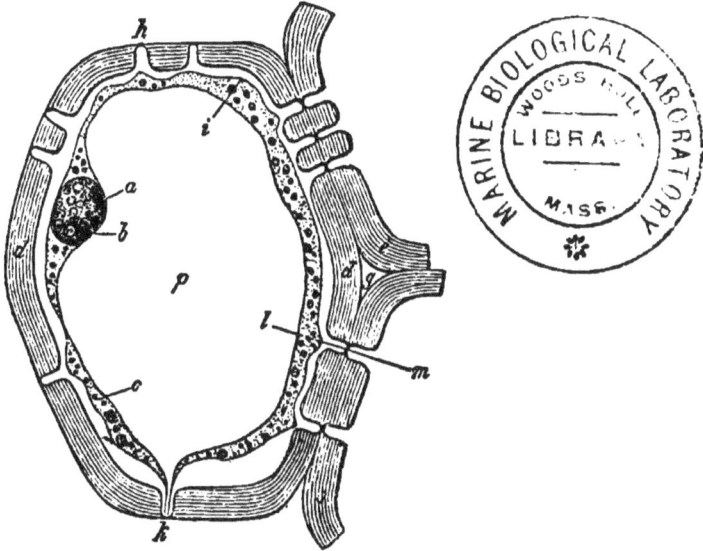

FIG. 4.

Cross section of pith cell of *Taxodium distichum*. *a* nucleus, *b* nucleoli, *c* protoplasmic body, *d* cell wall, *e* and *s* cell walls of two neighboring cells, *g* intercellular space, *h* pore canal, *i* outer layer of protoplasm, *k* simple pore of which the protoplasmic body has grown to the portion of wall separating this from the corresponding pore of the neighboring cell. Cell is represented in condition of plasmolysis. —(*Accd. to Th. Hartig.*)

so reduced in size that the protoplasm contracting around them becomes separated from the wall of the cell. In this condition the cell is said to be plasmolytic, and the process is known as plasmolysis. (See Fig. 4.) The living active cell is generally turgescent.

term turgor, as sometimes used for osmotical pressure, says : "As the extension of the cell wall and all the conditions resulting from the pressure of the cell contents on the wall are entirely independent of the cause or origin of this pressure, it is expedient to understand the entire sum of this tension under the term turgor."

1. Finer and Coarser Structure of Cell Wall.

THE wall is a product of the contents and is not an essential part of the cell. Its formation occurs in two ways. A naked cell may surround itself by a wall, or a cell already possessing a closed wall may form a new one which extends through the cavity and becomes fixed to the surfaces of the opposite walls, the mother or initial cell, as it is called, thus becoming divided into two cells. Examples of the former case may be found in swarm spores of various plants which either throw off their cilia or draw them in, and soon after surround themselves by a cellulose wall. The second manner of wall formation is best illustrated in the growing regions of ordinary plants.

Here the first appearance of the new wall on a cross-section, as seen by the aid of the strongest lenses, is that of a very fine line or thread stretched across the cavity. This is known to result from some action of the protoplasm of the mother cell by which cellulose micellae are formed and arranged in such a manner as to make a very thin membrane. This is the product either of the entire protoplasm of the original cell, or of that portion of it which is in near proximity to the newly formed wall. In either case it may be said to belong equally to both the cells resulting from its formation. At this stage of its existence it is known as the primary wall.

By watching its development it will be readily seen that it does not remain long in this condition, but increases in thickness till it reaches that of the other walls in the growing portion of the plant. The new cells thus formed now begin to increase rapidly in size and to assume those shapes which will fit them for their different functions. During this period the walls remain nearly uniform in thickness, their growth being limited to two directions, length and breadth, and this is known as surface growth. When the cell has reached its normal shape and size, by this process, another kind of growth may ensue, which results in a thickening of the wall. These two processes are generally referred to as growth in surface and growth in thickness, and are so sharply distinct from each other in reference to the time of their occurrence, that it is necessary to indicate which of the two is meant when speaking of the development of the wall.

The new material which is used in this development is the product of the protoplasm of the growing cells. There are two theories respecting the manner by which this is added to the primary wall. It is supposed to consist of cellulose micellae which either make their way between those of the primary wall and so incorporate themselves with it, or are added to its surface, becoming a part of it without penetrating between its micellae. The first method is known as intussusception; the second as apposition.

As a rule, surface growth is completed before that of thickness begins. In certain cells destined to carry on the various life processes of the plant, this subsequent growth in thickness never takes place. It is customary to speak of such cells as thin-walled, even though they vary greatly among themselves in this respect.

There is another class of cells whose principal function is that of the conduction of food and building material from place to place when it is necessary that these be carried considerable distances. Their walls are only partially thickened, so that liquid substances may easily pass through the thinner places, while at the same time the thickened portions furnish support and strength to the organ containing them.

Finally a third class is represented by the cells whose chief function is that of support. These are usually referred to as thick-walled in distinction from the first class. Their walls are more or less uniformly thickened, and comparatively few thin places occur.

It must not be inferred from this that the function of the cell can at once be determined by its shape and the nature of its wall. These are only indications which help us to decide its leading or principal function. The division of cells into three classes as here made, does not actually exist in nature, for each class passes into the next one by such imperceptible differences that it is impossible to say where one ends and the other begins. For purposes of study, however, it is convenient to group into classes, and this may be done in this instance by selecting the leading characteristics of the different cells.

There are certain facts in connection with the surface growth of newly formed cells, that lead to the opinion that the wall, at its very beginning, consists of two lamellae or plates, so closely connected that it is impossible even with the strongest lenses to see any indication of their line of contact. However this may be at first, as the cells grow older the double nature of the wall is seen by the separation at the corners by which the intercellular spaces are formed.

In many cells where growth in thickness has succeeded that of surface growth, this lamellated structure becomes very evident. For example, a cross-section through the thick-walled cells of pine wood shows three distinct layers in the walls. In cells of many other woods the existence of layers may be made evident by the use of various chemical reagents. In cork cells five distinct layers may frequently thus be made to appear. In such cases, the middle lamella is said to be common to each cell, while the other layers belong to the individual cell which they enclose.

The process of maceration dissolves the middle lamella of such cells, showing that there is a chemical difference between it and the other layers. In living, active, and moderately thin-

walled cells the existence of this lamella may be inferred from the fact that by maceration these cells also may be completely isolated from each other, each retain-
ing its own wall. The manner of origin of these layers is yet unknown. There are two possible ways of ex-planation. Either they exist from the first, but connect with each other in such a manner as to present the appearance of a homogeneous struct-ure, and afterward become distinct enough for the line of contact to be seen ; or they are the result of some chemical change occurring in the wall, which was at one time per-fectly homogeneous. (Fig. 5.)

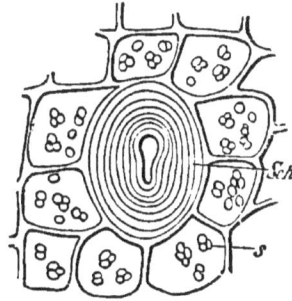

FIG. 5.

Tangential long section through the rind of cinnamon, used here to rep-resent stratification of wall. *Sch.* gives the appearance of stratified wall. *s.* starch grains.
(*Accd. to A. Vogl.*)

In bast cells of certain families this lamellated structure is connected with another which indicates the independent growth of each layer. The separate layers when seen in the cross-section appear to be divided into narrow strips or blocks by lines crossing the layers more or less obliquely. (Fig. 6.) The surface of such cells, when exam-ined in a longitudinal section, is seen to be crossed by spiral lines generally running from right to left. (Fig. 7.) To the structure indicated by this ap-pearance has been given the name stria-tion, or striping, while the lamellated structure is called stratification. Any one of these lamellae, if considered by itself as a hollow cylinder, appears to

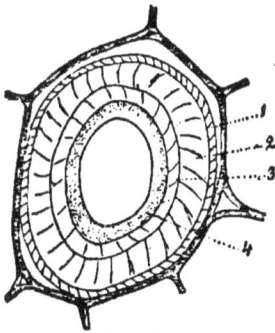

FIG. 6.

Cell showing stratified wall, the different layers of which are crossed by lines denoting stria-tion. — (*G. Krabbe.*)

be made up of a number of spirally wound strips which are distinct enough from each other to show their line of contact.

In these few instances it has been discovered that the separate lamellae are added, as such, to the primary wall. Each layer is formed in some way out of the protoplasm and joined to the old wall. Subsequent growth in thickness of each or any layer may take place, so that this furnishes no evidence in

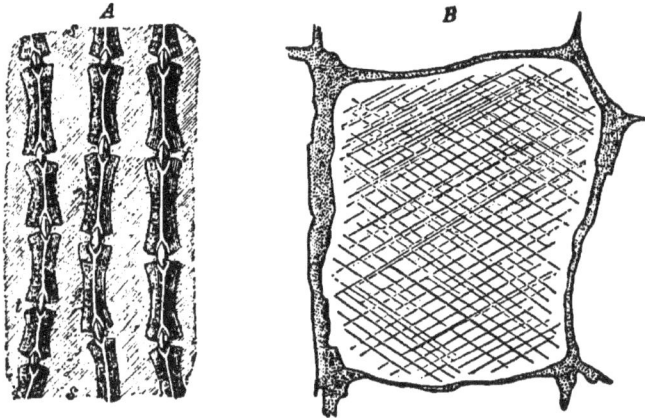

FIG. 7.

A Fragments of tracheids of the fir in tangential long section. *t* bordered pores. *s* striation of wall. *B* cell from the parenchyma of *Dahlia* tuber showing striation. × 240.
(*Accd. to Strasburger.*)

regard to the question of growth by intussusception or apposition. Both of these methods of growth are possible here as in most other cases which have been studied. There is much difference of opinion in regard to these two theories, though it is now generally believed that both methods of growth occur.

The structure of wall described as striation is not limited to the above-mentioned instances, but is often found in cells of various kinds. Neither is it always connected with stratification, being often found in walls which show no appearance of division into layers.

It has already been stated that there are two classes of so-called thick-walled cells, those designed principally for support and those both for conduction and support. In the former class

the new material is added in such a way that the completed
wall is nearly uniform in thickness. On a cross-section through
such a cell the limiting surfaces of the wall appear bounded by
straight lines. (Fig. 8.) In the second class the walls are
thickened irregularly, the new material being added only to
certain portions, leaving thin places through which the liquid
contents may pass easily and rapidly from cell to cell. This
explains the use of the terms "finer" and "coarser structure"
of cell wall. By "finer
structure" is meant
stratification or stria-
tion or both. "Coarser
structure" refers to the
more evident sculptur-
ing seen on the walls
of the latter class.

In reference to its
direction, growth in
thickness may be either
centripetal — toward
the centre of the cell
— or centrifugal —
from the centre. In

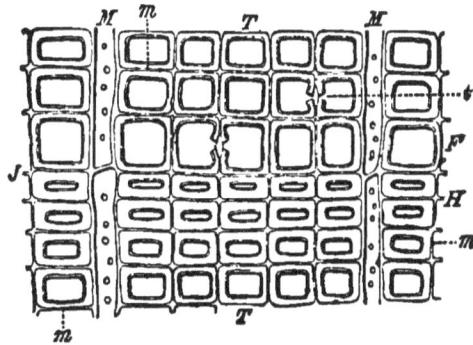

FIG. 8.

Cross section through the wood of *Abies pectinata*. *T T*
tracheids, *J* year's ring, *H* fall growth, *F* spring
growth, *t* bordered pore, *M* medullary ray, *m* middle
lamella. × 300. — (*Wiesner.*)

the latter case it must occur on that part of the wall not in
connection with the protoplasm. It is evident that centrifugal
growth, in this sense, can occur only in cells originally separated from each other, and in those first connected by common
walls which have later become partially separated by the
division or splitting of their walls.

The best example of centrifugal thickening of separate
cells is the pollen grain, the outer surface of whose walls is
often covered with projections of various forms. Among cells
still partly united, examples occur where parts of certain cells
of a tissue grow out into neighboring intercellular spaces.

Centripetal growth is by far the more frequent and includes all growth in thickness of walls common to two cells. It is found in cells of both classes, those where the thickening is uniform, and those where it is irregular. Of the former class, the ordinary bast and libriform cells furnish the best examples, where the primary wall is increased in thickness centripetally until only a small cavity remains.

Of the latter class there are various kinds. Beginning with those where only a small portion is thickened, the larger part remaining thin, we have:

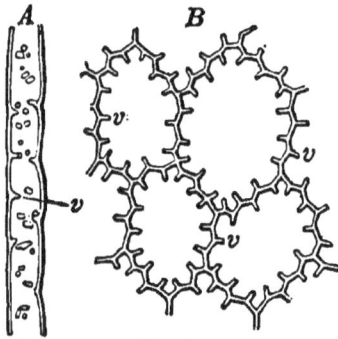

FIG. 9.

A piece of rhizoid of *Marchantia polymorpha*. *B* epithel cells of petal of *Pelargonium zonale*. *r* projections into cell lumen. × 300. — (*Wiesner.*)

I. Small circular spots of surface thickened, producing cone or rod-like projections which extend into the lumen.

EXAMPLES: Rhizoids of Marchantia and certain ducts whose walls are strengthened by conical inward projections. (See Fig. 9.)

II. The thickening is not limited to spots but runs in continuous spirals with varying closeness of coil.

EXAMPLES: Spiral ducts.

III. Wall thickened in separate rings.

EXAMPLES: Ring ducts. (Fig. 10.)

IV. The thickened ridges run parallel to the long axis of growth and are connected by short cross-bars.

EXAMPLES: Ladder, or scalariform ducts.

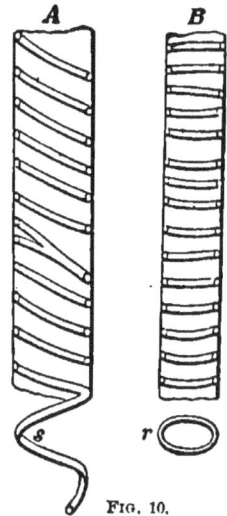

FIG. 10.

Fragments of ducts from stem of Rye in long section. *A* spiral. *B* ring duct. *r* ring loosened and separated from duct. × 250. — (*Wiesner.*)

V. A form similar to the fourth, but the scalariform thick-enings are somewhat irregular.

EXAMPLES: Reticulated ducts.

VI. A modification of the fifth, but the thickened portion of surface is greater than that left thin, and the small thin portions are either circular or lens-shaped.

EXAMPLES: Ducts with simple and bordered pores.

Another class of cells, designed both for conduction of material and for support to the organ containing them, is

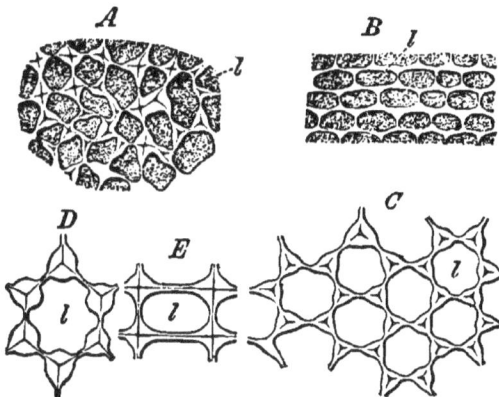

FIG. 11.

Cross-sections showing Collenchyma. *A* from stem of *Chenopodium album*. *B* from stem of *Sambucus nigra*. *C* from petiole of leaf of *Nicotiana macrophylla*. *D* from stem of *Ceratophyllum demersum*. *E* from petiole of leaf of *Phytolacca decandra*. With the exception of *B*, the outer limits of the common walls may be seen. *A* and *B* × 200. *C* and *E* × 400. — (*Accd. to Wiesner.*)

known as collenchyma. Cells of this class are generally longer in one direction than in the other two, and the thick-ening takes place along the corners or angles, leaving a narrow portion free for the transmission of contents. Such cells are found especially in portions of plants whose growth in length is not yet finished, while the six classes previously described belong to those parts where growth in length is ended.

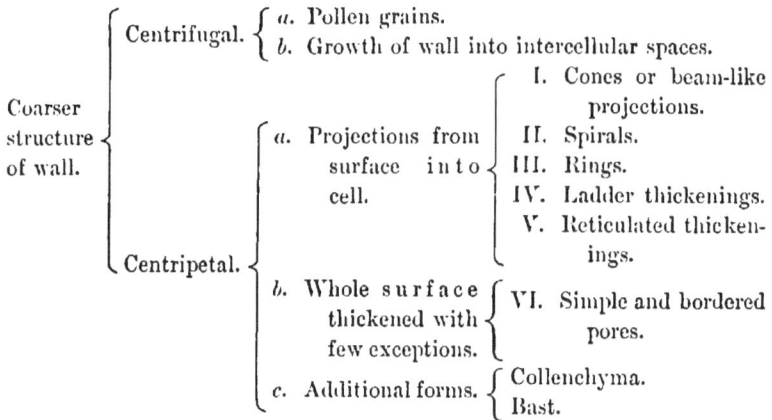

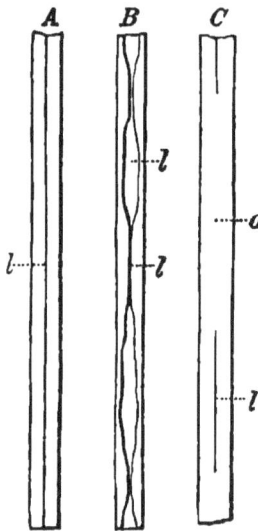

Coarser
structure
of wall.

- Centrifugal.
  - *a.* Pollen grains.
  - *b.* Growth of wall into intercellular spaces.
- Centripetal.
  - *a.* Projections from surface into cell.
    - I. Cones or beam-like projections.
    - II. Spirals.
    - III. Rings.
    - IV. Ladder thickenings.
    - V. Reticulated thickenings.
  - *b.* Whole surface thickened with few exceptions.
    - VI. Simple and bordered pores.
  - *c.* Additional forms.
    - Collenchyma.
    - Bast.

FIG. 12.

Fragments of isolated bast cells. *l* lumen of cells, *A* from *Linum usitatissimum*, *B* from *Corchorus capsularis* irregularly thickened, *C* from *Sponia Wightii*, lumen in some places entirely filled with thickening material. × 300. — (*Wiesner*.)

Of these various forms, those with pores require special notice. The simple pore results from a small portion of the primary wall being left unthickened on both sides. In this way a small channel is formed with a membrane across its central portion, the pore not being a complete opening between the two cells as its name implies.

The bordered pore is a complex little piece of mechanism, varying somewhat in different cells and also in different plants. The portion of primary wall which is left unthickened is here usually larger than in case of the simple pore. Instead of the material being added so as to form a single channel, the thickening portion gradually extends itself over that part of the primary membrane left free, in the form of an arching roof, but never entirely covers it. An aperture is always left, and the new material is so deposited about this opening as to form another channel similar to that of the simple pore. Thus there are two distinct parts to the opening, called the court and canal. The court is also called the limb or border, hence the name bordered pore. As in the simple pore, this structure has an exact counterpart, opposite it, in the adjacent cell. and the two together make up the complete pore which belongs equally to both cells. The thin portion of membrane between the two parts is called, in German, *Schliesshaut*, or closing membrane. This name comes from its supposed function in cases of large and well developed pores like those of pine wood. Here its central portion is slightly thickened, a narrow border around this remaining thin and flexible. If water be pressed into cells containing these pores it passes through the opening, presses against the Schliesshaut or disk, the flexible portion of which yields, so that the whole membrane is pressed back against the aperture of the opposite part of the pore, the thickened portion lying over the opening. In this way the capacity of one cell is increased at the expense of the others.

FIG. 13.

Bordered pores as seen on edges of wall and on surface. *a* pore with slit-formed canal and strongly thickened closing disk. *b* pore with cylindrical canal; the closing disk here rests on the wall of the court. *c* pore with oval canal; the disk has grown fast to the wall. (*Th. Hartig.*)

2. Chemical and Physical Properties of Wall.

The wall consists of solid matter and water. If the latter be driven away by evaporation the remainder is found to consist of two sets of substances, organic and inorganic. This is found by burning the dried wall, whereupon the organic matter passes away in the atmosphere, while the inorganic remains

behind, constituting the ash. These mineral matters are calcium, sodium, potassium, magnesium, iron, silicon, phosphorus,
sulphur, chlorine, and some others. The exact nature of their
union with the organic substances is not known. It is supposed, however, to resemble that of alloy among metals, rather

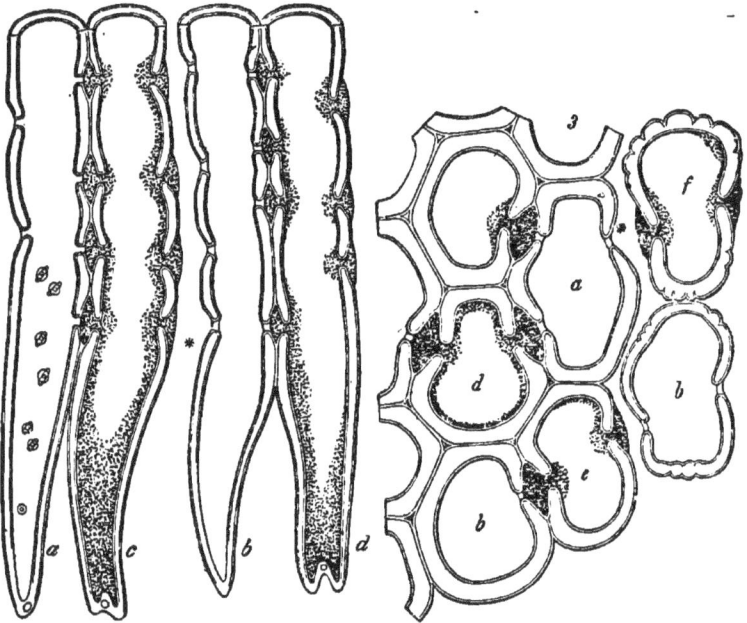

Fig. 14.

Cells with bordered pores seen in long section to show how the plate prevents the contents
of one cell from passing into the neighboring one. The dark portion represents
contents colored with carmine. — (*Th. Hartig.*)

than a true chemical union. In some instances the mineral
matter comprises a separate portion of the wall, as the silica in
Equisetum, the calcium which forms part of the cystoliths found
in certain cell walls, also that which occurs on the surface of
some plants, as in the Characeae, one or two families of the
Florideae, and different species of Potamogeton.

By far the greater portion of solid matter comprising the cell wall consists of cellulose, a substance of the same formula as starch, $C_6H_{10}O_5$. It resists the action of most solvents, yielding only to concentrated sulphuric acid, to certain ferments, and to copper oxide of ammonia. When treated with iodine and followed by weak sulphuric acid it turns blue, and with chlor-iodide of zinc it turns violet. The young walls of all plants consist largely of pure cellulose. With age, however, certain modifications take place in the cellulose of some cells which are designed for special functions ; the principal changes are the following : Lignification ; Suberization : Partial or complete conversion into mucilage, Gelatination.

3. Lignification of Wall.

Lignin, or the woody portion of plants, consists of the same elements as cellulose, but with carbon largely predominating. The process by which cellulose is changed to lignin is not known. There are some grounds for supposing that lignin is a substance formed by the activity of the living protoplasm of the wood cell, and deposited among the cellulose micellae as an incrustation, which works a chemical change in them. In the walls of mature wood cells there are generally found three layers, primary, secondary, and tertiary ; the primary lying next to the middle lamella, and the tertiary being in contact with the protoplasm. The amount of lignin in these layers varies, the primary generally containing more than the other two.

4. Suberization and Cutinization of Wall.

The walls of those cells which are designed to protect the plant from outward injury, such as too rapid evaporation and extremes of heat and cold, undergo a change by which the cellulose becomes suberin and cutin. These substances are very similar in their chemical nature, and are both very rich in

carbon. Of their manner of origin and connection with the
rest of the wall, still less is known than in the case of lignin.
The outer walls of epidermal cells are generally cutinized, sometimes only the outer portion, and sometimes the entire wall, and
in some instances the cutin extends throughout all the epidermal
walls. Suberin is found in the walls of the periderm cells which
are known as cork, or cork cells. Of the three layers of these
walls, the primary seldom contains any suberin but is often
lignified, the second is suberized, the third contains but little
suberin, and in many instances is found to consist of pure
cellulose.

5.   Formation of Mucilage and Gums.

The third change, or breaking down of cellulose into mucilage, is caused by the swelling of the membrane on the entrance
of water.   In higher plants, this occurs frequently in the
epidermal cells, as in flax-seed; here the cellulose part of the
epidermal wall swells till the outer cutinized portion bursts
off and the entire layer of cells is reduced to a mucilage.   In
lower plants one or more of the outside layers of the walls
swell, so that a mucilaginous coating is formed about the plant.
The formation of various gums is a particular phase of this
change of cellulose and requires a special chapter for treatment.

The most prominent physical characteristics of cell membrane have been mentioned in the preceding description.   Its
power of double refraction has been taken as evidence of the
crystalline nature of the individual micella or unit of structure.
It possesses the quality of hardness in varying degrees, depending upon the amount of mineral matter in its construction.   It
is elastic as well as flexible, being able to assume its normal
size after it has been stretched beyond this by turgescence.
It is capable of imbibition, and is in nearly all cases permeable
by liquids and gases, though the process is a slow one.

CHAPTER III. — CELL CONTENTS.

IN active cells there are generally a number of structures sharply differentiated from the ground substance. To the latter Strasburger has given the name *cytoplasm*, in order to distinguish between protoplasm in its undifferentiated state, and that containing the various products formed in and by it. The structures are divided into two classes, active and passive; the former are similar in nature to cytoplasm, and are the nucleus, chlorophyll grains, color grains, and starch-builders, the last three being also known as plastids. The passive structures are those which may be formed at any time in the cell, and in comparison with the former class are less active in carrying on the life processes. These are protein or aleurone grains, starch grains, globoids, and crystals. In addition to these definitely organized structures, there is another large class, no less important but of less definite structure, which may be termed organic in distinction from organized. For convenience of description the entire cell contents may be tabulated as follows :

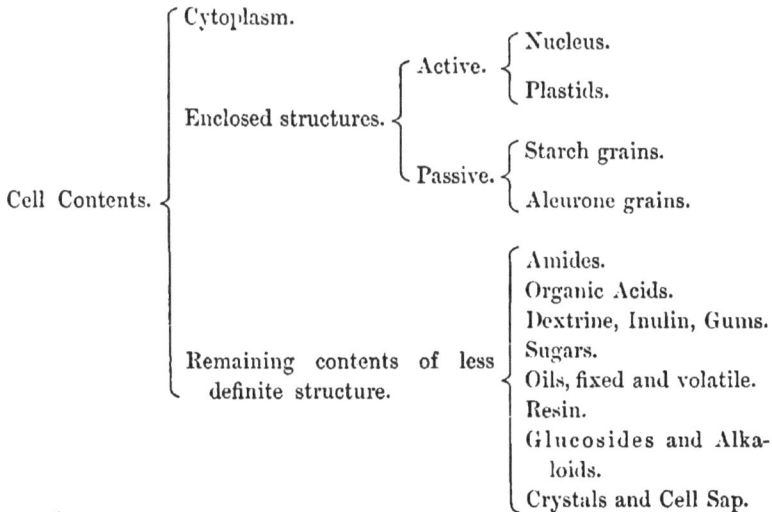

Cell Contents.
- Cytoplasm.
- Enclosed structures.
  - Active.
    - Nucleus.
    - Plastids.
  - Passive.
    - Starch grains.
    - Aleurone grains.
- Remaining contents of less definite structure.
  - Amides.
  - Organic Acids.
  - Dextrine, Inulin, Gums.
  - Sugars.
  - Oils, fixed and volatile.
  - Resin.
  - Glucosides and Alkaloids.
  - Crystals and Cell Sap.

1. Cytoplasm.

When closely examined this appears to contain fine grains, much smaller than the structures just described and having the nature of proteids. They are imbedded in a clear, hyaline substance, and are unequally distributed, occurring in greater numbers near the central part of the cytoplasm; the outer part or that lying near the cell wall is nearly clear. Some botanists believe that this granulated appearance of the cytoplasm indicates a definite structure similar to that known to exist in the animal cell. They claim that the apparent grains are really not grains, but very slender fibers running through the hyaline substance with some fixed order of arrangement. No definite proof of this has yet been found. According to the amount of water contained in it, the cytoplasm forms a more or less tenacious, mucilaginous mass in which the single particles or granules may easily move. The chief constituents of this mass are known to be of proteid nature, but their exact chemical composition has not yet been determined. The elements are carbon, hydrogen, oxygen, nitrogen, and sulphur and phosphorus, the last two in small quantities, but in what proportions their atoms unite to form a molecule is not known. The many experiments that have been tried to determine this have succeeded only in establishing a probability that very complex chemical compounds are present; thus rendering possible a large number of different organic combinations on the dissociation of these compounds.

This mass of cytoplasmic substance is flexible, slightly elastic, and rich in colloidal substances which in order to become diffusible must be changed into other nitrogenous compounds named amides, of which mention will again be made. When living, it is almost impenetrable for salts and coloring matters, but when dead this property is lost.

The power of motion residing in vegetable protoplasm is of

two kinds ; first, that which is limited by the cell wall, and which produces no change in the position of the cell or plant ; second, that by which the plant is enabled to move from place to place. In certain cells currents of protoplasm have been discovered varying in their action according to the distribution of vacuoles and consequent arrangement of the protoplasmic substance. Where there are numerous vacuoles extending more or less evenly through the cavity, the current is irregular, following the windings of the protoplasmic meshes. In such cases it is called circulation. An example of it is found in stamen hairs of Tradescantia. In other cells where the greater part of the protoplasm is pressed back upon the cell wall, a regular current is seen passing entirely around the cell, carrying with it some of the structures before referred to. This action is known as rotation. Example, Nitella.

The second kind of motion occurs in case of small, unicellular plants with or without membrane, and corresponds exactly to that in animals of like low organization. Among certain bacteria, and also swarm spores of various algae and fungi, cilia occur, by whose alternate contraction and relaxation the cell is enabled to move about from place to place. There are also small multicellular plants having the power of motion from place to place, the mechanism of which is supposed to be similar to that producing certain motions in the members of higher plants, where the individual as a whole occupies a fixed position.

2. Cell Nucleus.

This appears as a roundish oval or lens-shaped body sharply set off from the surrounding medium ; it consists of a clear substance containing numerous small granules and one or more larger grains called nucleoli. It is bounded by a dense portion of the clear ground substance. Cells ordinarily have but one nucleus, but in many of the lower plants several nuclei occur in the same cell.

In cormophytes the presence of one nucleus has been traced in nearly all the living cells. These nuclei are nearly spherical in form, their outer portion much denser than the surrounding cytoplasm, while the inner is only slightly so. They are largely albuminoid in nature. In the cells of the sexual organs in certain stages several nuclei are normally present, and in cells of purely vegetative organs a number of nuclei are often found after an injury.

In many of the thallophytes until quite recently there was no definite proof of the existence of nuclei. Now it is claimed that by a more successful use of clearing and staining agents, nuclei have been found in many algae, in some instances singly in a cell, in others in numbers. In the mycelium of many fungi a plurality of nuclei has also been determined.

So far as has been ascertained, the chemical nature of the nucleus is the same as that of the surrounding cytoplasm. Its physiology is not clearly understood. Formerly it was supposed to be intimately connected with the process of cell-division. It is now known that the two processes, cell-division and nuclear division, are not dependent upon each other. Some scientists claim that the function of the nucleus is to aid in the production of proteid substances, others consider it the medium of inheritance, or transmission of characteristics to the succeeding generation.

The number of nuclei is increased only by the division of those already in the cell. This division is of two kinds, direct, or *fragmentation*, and indirect, or *karyokinesis*.

3. Direct Division of Nucleus.

This method of division is not preceded by any special changes in the constitution of the nucleus, nor is it followed by cell-division. In the higher forms of plants there is very good evidence that it occurs, but only in a limited number of cases ;

among lower plants it has only been traced with certainty in the older cells of the Characeae. The process consists in the elongation of the nucleus, the central portion becoming small or thread-like, and the two halves stretching apart until the thread breaks and the two parts both assume the regular form of the nucleus. Such a process has been observed in the cells of Tradescantia and in the bast cells of *Urtica urens*.

4. Indirect Nuclear Division, or Karyokinesis.

The different phases through which the nucleus passes in this method of division has been studied in various plants. As some stages appear much more clearly in certain plants than in others, the descriptions of the process of karyokinesis in the plant cell are generally taken from several sources and combined in order to illustrate the entire process. For the first steps, the nuclear division in the embryo-sack of *Fritillaria imperialis* furnishes a good illustration. Pollen mother-cells of different plants show better the formation of the cellulose wall. This is usually considered the final step in the process of karyokinesis, but it is not true that indirect nuclear division is always accompanied by cell division.

Nuclei, when about to divide, are large, well-defined bodies possessing either a single nucleolus or several, and containing also a substance distinguished from the remaining contents by its form, which is either that of small distinct granules scattered in the cytoplasm, or that of a combination of these granules in such a way as to form a network of fine threads. This is the part of the nucleus known as chromatin, or that which takes the dark color when staining fluids are used. The first appearance of division in Fritillaria is the formation of this substance into a number of short thick threads, called nuclear segments. In other cases, these segments are simply irregular masses of less definite shape. (See Fig. 15, 2.) Shortly after this for-

mation, the membrane of the nucleus appears to be removed or
dissolved, the nucleoli disappear, and the segments arrange
themselves near the central portion of the nucleus.   Next
appears the so-called spindle, composed of very fine fibers or
threads, which run through the segments and the entire length
of the nucleus, converging at the opposite ends, which are called
poles.   The substance of these threads is of a different nature
from that of the segments, as they either remain uncolored by
the use of staining agents or take a much lighter color.   The

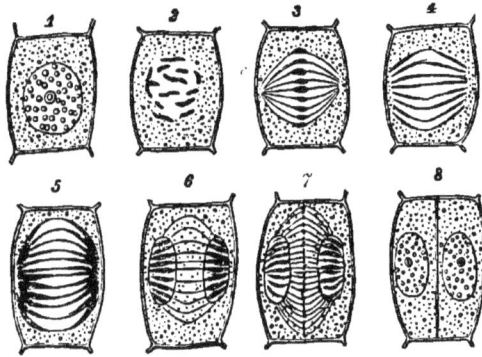

FIG. 15.

Division of nucleus and cell of the mother cell of a stoma. — (*Accd. to Strasburger.*)

segments lying in the central portion of the nucleus form what
is known as the *equatorial* or *nuclear plate*.   In Fritillaria they
are in the form of bands or flattened rods; each band now
divides lengthwise, forming twice the original number.   These
daughter-segments then separate and move toward the poles,
half to one and half to the other.   There they form the founda-
tion of the two new nuclei.   They gradually change their form
and break up in separate portions, which unite to form at last a
network similar to that of the original nucleus.   Shortly after
this, a nucleolus appears in each, and a firm membrane surrounds them and the process of nuclear division is complete.

Essentially the same process occurs in other cases where the segments are less regular in shape and number than in Fritillaria. This is illustrated by Figs. 4–7, in Plate 15.

During this time the spindle fibers retain their shape and size. About the time of the completion of the two nuclei, a little swelling appears near the middle point of each of these threads, so arranged as to form a plate through the equatorial plane, or the place formerly occupied by the nuclear plate. These swellings increase in size until they finally coalesce and form what Strasburger has named the *cell plate*. This plate is intimately connected with the formation of the cellulose wall which soon afterward appears, dividing the cell into two daughter-cells.

It has been found very difficult to ascertain the exact numerical relations between the spindle fibers and the segments. It is possible, however, that there are as many fibers as segments, and the latter are supposed to move along the fibers to the poles. There are also some grounds for supposing that the substance composing the spindle fibers is not a part of the original nucleus, but of the cytoplasm of the cell. For example, it is said that these fibers never appear until after the membrane of the nucleus is dissolved, and that previous to the disappearance of this membrane, the nuclear segments are the only differentiated structures found in the substance of the nucleus.

5. Plastids.

Plastids are of three kinds, chloro-, chromo-, and leucoplastids. The first includes all structures which contain chlorophyll or the green coloring matter of plants. This is a pigment found in all classes of plants except fungi, and generally connected with some certain portion of the protoplasm. In some of the low algae, however, it is equally distributed through the whole mass of protoplasm. It is also claimed that it is occasionally found in other plants, not connected with any particular structure, but this is supposed to be a pathological rather than normal appearance.

The chloroplastid, then, consists of a protoplasmic framework and a green pigment. In phanerogams, vascular cryptogams, and nearly all Bryophytes, this framework takes the form of a small spherical or egg-shaped body. In one of the groups of Bryophytes, Anthoceroteae, only one large plastid occurs in a cell. In the algae plastids occur in various forms, such as spiral bands, rings, stars, and other singular shapes.

The pigment can be abstracted by the use of absolute alcohol, leaving the framework behind as a colorless, somewhat porous substance, having the same chemical nature as the remaining protoplasm. If benzol be added to the solution of the pigment in alcohol, a yellow substance is separated out which remains in the alcohol, the green of the chlorophyll going over to the benzol. This yellow substance has been named Xanthophyll. It is present in the chloroplastid, but whether in addition to the chlorophyll itself, or forming an integral part of the pigment, is not known.

An alcoholic solution of this chlorophyll pigment has the quality of fluorescence. When seen in reflected light it appears opaque and of a deep red color. If the light which has been passed through a layer of a moderately strong solution be examined with the spectroscope, characteristic absorption bands will be seen. Chemically considered the pigment is a combination of carbon, hydrogen, oxygen, and nitrogen, with no iron in it, but it is never formed unless iron be present in the tissues. With some few exceptions its production is also dependent upon light. When an insufficient amount of light is present, a substance named etiolin is formed, which in color resembles Xanthophyll, but is not supposed to be identical with it.

In the chloroplastid, in connection with the action of light, a carbohydrate originates, probably a formic aldehyde, which soon changes to sugar, and when this material is not carried off fast enough to prevent its aggregation in the chlorophyll grain, it is again changed into starch. This process is known as CO_2 assimilation, as it begins with the separation of CO_2 and H_2O into their elements, which are afterward reunited into carbohydrates. In the Florideae and other algae with different colored pigments the chlorophyll exists but is concealed by the other coloring matter.

Chromoplastids include all color grains except the chloroplastids. In the higher plants they are not confined to the flower leaves but occur in various parts. Very few have been found in the lower plants; in vascular cryptogams they occur only in the fertile stem of *Equisetum arvense*, and in mosses, in the antheridium wall. In form they vary from small, nearly spherical, to long, rod-shaped bodies. While they furnish the coloring matter of plants to some extent, in most cases the color pigment is not connected with any structure but is dissolved in the cell sap.

Leucoplastids are small colorless structures whose function is still a disputed question. They were first discovered in 1854, but little notice was taken of them until Schimper took up the subject in 1880. They are very easily dissolved, and for this reason were so long overlooked. They are found in nearly all cormophytes but rarely in thallophytes. They occur in stems which contain reserve material, in tubers, and in growing parts of plants where cells are dividing, also in epidermal cells. They vary in form from nearly spherical to flat, lens-shaped bodies. It has been proved that the starch grain originates in the chloroplastids, also that after it changes into a soluble carbohydrate it again appears in cells not containing chlorophyll. In these cells not only are leucoplastids present, but they are often found with starch grains in various stages of development imbedded in them or adhering to their surface. From these facts some have drawn the conclusion that the office of the leucoplastid is that of a starch-builder, and that whenever the smallest particles of starch are formed into an organized structure it is done through the medium of this plastid.

Regarding the origin of the plastids there are two views, one, that they are formed by the cytoplasm at any time and whenever needed, the other that they increase in numbers only by division. Those holding the latter view claim that certain structures exist in the fertilized egg cell, such as nucleus and leucoplastid, and that as new cells are formed these structures divide to supply them, and that the colorless leucoplastids may later be impregnated with pigment so as to form both chloro- and chromoplastids.

6. Starch Grains.

It is customary to divide the food products in plant cells into two classes, nitrogenous and non-nitrogenous. The organized structures of the latter take the form of starch grains, which originate in the chloroplastid by the process of CO_2 assimilation and are changed into sugar before they can be transported from the place of origin. Sometimes in the course of the passage of sugar from cell to cell, starch occurs as very fine grains. In this form it is called *transitory*, as it changes rapidly to sugar and this again to starch. In cells destined to hold food for future use, starch is laid up in large grains. It is then called *reserve starch*. It is in the formation of these grains that the ieucoplastids are supposed to play an active part, though it is not proved that they are not sometimes formed without the aid of these structures.

The form of the starch grains is generally roundish, elliptical, or egg-shaped. When crowded together in a cell they may become polyhedral. In the milk tubes of Euphorbiaceae they are club-shaped with enlarged ends.

The substance of the starch grain is closely related to cellulose, and has the same formula, $C_6H_{10}O_5$. The grains have ·

the property of swelling to a paste in hot water, and they turn blue on the application of iodine. A certain portion only takes this blue color ; it is called granulose, and dissolves more readily than the remaining part. The latter, called starch cellulose, turns yellow with iodine. The normal means of disintegrating the starch grain is by a ferment named diastase, which occurs in those cells where reserve starch is laid up. The granulose may be eliminated artificially by various weak acids ; also by saliva from the mouth when heated, leaving the starch cellulose as a skeleton. It is now claimed by some authorities that there is but one substance in the starch grain, namely, granulose. They hold that the starch cellulose, which they call amylodextrine, does not exist until produced by the action of the ferment.

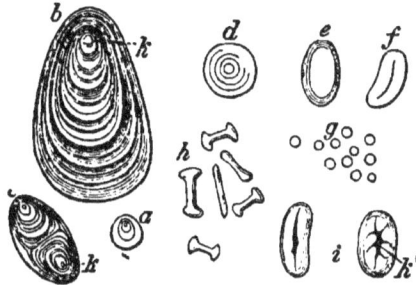

Fig. 16.

Starch grains. *a–c* from potato ; *c* young, simple grain fully grown ; *c* half-compound. *d f* large, *g* small grains from endosperm of wheat seed ; *d* after treatment with chromic acid. *h* from milk tubes of Euphorbia ; *i* from bean ; *k'* air space.
(*Wiesner.*)

The larger grains show a lamellated structure, the layers extending around a kernel or nucleus, whose position is either central, or more or less eccentric. There may be more than one nucleus in a grain, and in respect to the number of nuclei and the arrangement of the layers about them grains are divided into three classes, simple, compound, and semi-compound. The simple grain contains but one nucleus, the compound more than one, each being surrounded by its corresponding layers. The semi-compound grain has also more than one nucleus, but in addition to the layers surrounding each nucleus there are also layers enclosing the whole.

Naegeli, in his well known work on the starch grain, was

the first to give a satisfactory explanation of the origin and
nature of these layers. He claimed that the appearance of
stratification was caused by a difference in the amount of water
in the alternating layers. There are many facts which support
this view. One of them is that if the water of the grain is made
to evaporate the lamellated structure disappears. Since the
time when Naegeli advanced this theory, other facts have been
discovered which lead to the conclusion that there is no difference in the amount of water contained in the different layers,
and that the dark lines do not represent layers, but merely lines
of contact between the separate lamellae. This same reasoning
is also applied to the stratification of the cell wall.

7. Aleurone Grains.

While starch represents the organized form of non-nitrogenous reserve food, the aleurone grain is the organized form
of the nitrogenous material. These grains were discovered by
Theodore Hartig in 1855. They occur in great abundance in
those seeds which contain oil instead of starch. They are mostly
round or oval colorless bodies, though in certain seeds they are
colored. For example, they are yellow in Ailanthus, blue in
Zea Mays, green in Pistaciae. Their size varies in different
plants, and also in the same plants and cell. The largest grains
are found in cells containing little starch. In the seed of Vitis
there is a single large grain in a cell.

The grain itself consists of an outer covering of proteid
matter, which may enclose three different structures : protein
crystalloids, globoids, and real crystals. The crystalloids differ
from crystals by the property of swelling when water is added.
They are insoluble in water, but soluble in weak potash. As
their name implies, they consist of proteids, compounds whose
chemical nature is not yet thoroughly understood. The globoids
are nearly spherical in shape, and consist of salts of calcium and

magnesium, phosphoric acid, and a little organic matter. They are widely distributed, occurring in nearly all aleurone grains. The real crystals consist of calcium oxalate; they are of less

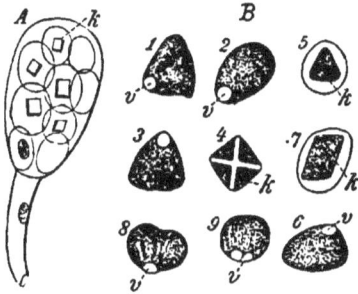

FIG. 17.

A club-shaped hair of an etiolated potato stem with crystalloids, k. B aleurone grains 1-5 from Ricinus seed; 8-9 from Momordica seed. r r globoids. k k crystalloids. A × 300. B × 400. — (Heed. to Vogl.)

FIG. 18.

Aleurone grains from Ricinus, a in thick glycerine, b in thin, showing the enclosed crystalloids. The roundish bodies are globoids. c grain containing crystal of calcium oxalate.

(Th. Hartig.)

frequent occurrence than the globoids, and are usually either needle-shaped or prismatic.

In some instances the entire proteid matter of the aleurone grain takes the form of a crystal and is then known as a protein crystal instead of a protein grain. Between these protein crystals and the grains with several enclosed structures, various intermediate or transition forms occur.

Reserve material is also stored in various places in the plant in a different form from the two organized structures, starch and aleurone grains. For example, the carbohydrates are sometimes stored as cellulose, the nitrogenous substances, as amorphous proteids. All of these substances are non-diffusible and must undergo some change when it becomes necessary to convey them from one cell to another. The non-nitrogenous substances are changed to sugars, the nitrogenous to amides.

This brings us to the consideration of the remaining substances found in plant cells, many of which are organic but not

organized. These have been referred to in the table as substances of less definite structure.

8. Amides.

These are soluble compounds, capable of crystallization, and
are found in greater or less quantities in all young growing
parts of plants. As already stated they arise from the dissociation of proteids and protoplasm, in order to make their
transfer possible, and it is claimed that they are also formed by
synthetical processes. The most widely distributed amide is
asparagin, which is found not only in the young shoots of as-
paragus but in nearly all plants. Leucin, tyrosin, glutamin,
and others occur less frequently.

9. Sugars.

These are widely distributed in the different plant tissues,
and are divided into two groups, cane and grape sugars. The
latter have the formula $C_6H_{12}O_6$ and are fermentable. The
principal sugars of this group are dextrose or glucose, and
laevulose or fruit sugar. Glucose is the most common form,
and is the one in which the most of the non-nitrogenous elements
of food and building material are transported from place to place
in the plant.

The formula of the cane-sugars is $C_6H_{22}O_{11}$, and they are for
the most part unfermentable. One of this group, known as
saccharose, occurs as reserve material in such plants as beets,
sugar-canes, and maples. This, in order to be transported from
the cells where it has been stored, must be changed to grape-
sugar. Another important kind is maltose, the sugar formed by
the germination of barley seeds.

All, or nearly all, of the non-nitrogenous food products contain the three elements, carbon, oxygen, and hydrogen, and in
works on plant physiology they are generally divided into two

classes. Those containing oxygen and hydrogen in the same proportion as in water are called carbohydrates. There are various other products which contain the same elements but in different proportions. The principal group is that known as oils, but there is one class of these which lacks oxygen.

10. Fixed Oils.

These are combinations of fatty acids with glycerine. According to the kind of acid they are either liquid or solid. They are probably always present in protoplasm in very minute particles ; and often as drops of considerable size, which are in some cases so numerous as to nearly fill the entire cell cavity. This is especially true of the fungi, in whose cells fat takes the place of starch. Wax is a form of solid fat which is not a reserve material, but a secretion in the outer wall of the epidermis, thrown to the surface in the form of small grains, rods, or scales. It also occurs in very fine particles on the surface of certain fruits. forming the so-called bloom.

11. Volatile Oils.

These differ from the fixed in possessing a volatile or ethereal quality by which they give out a pungent odor. They probably never occur as reserve materials, but as secretions which are either held suspended in the cell sap, or thrown out into intercellular spaces, or into peculiar cells designed to receive such secretions. Several of the volatile oils lack oxygen, — for example. turpentine, whose formula is $C_{10}H_{16}$. A few others contain not only the three elements carbon, oxygen, and hydrogen but also a little sulphur.

12. Resins.

These are closely related to the ethereal oils, some being regarded as their oxidized products. They are widely distributed and are divided into three classes, true, balsam, and gum resins, the balsam being a mixture of resins and ethereal oils, together with aromatic acids. Gum resins are a mixture of resins, gums, and ethereal oils. They may originate in cells from which they are excreted into intercellular spaces, or they may arise from the breaking down of the cell wall and are thus often found mixed with unchanged parts of plant tissue. Caoutchouc is a substance related to resins and originates in certain cells called milk tubes where it forms an emulsion with the water of the sap.

13.   Dextrin, Inulin and Gums.

Dextrin is a substance very closely allied to starch and is found dissolved in the sap of many plants. Sinistrin and Inulin

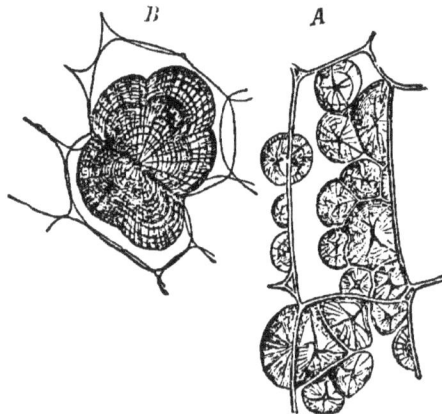

FIG. 19.

Inulin crystals. *A* from tuber of *Dahlia variabilis*. *B* from *Helianthus tuberosus*. Crystals appear after the material has been treated with strong alcohol. × 500.—(*Accd. to Sachs.*)

also occur in solution in sap, the latter in considerable quantities in the underground parts of the order Compositae, where it takes the place of starch. It may be precipitated as crystals of characteristic form, by the use of alcohol. There are several kinds of gums, some of which arise from the breaking down of the cellulose of the wall, others as secretions which are developed in certain cells designed for this purpose.

14. Organic Acids.

Oxalic acid is found, like many other acids, free in the cell sap, but is most frequent in combination with calcium. As calcium oxalate, it is found in the various kinds of crystals occurring both in walls and in contents. Calcium oxalate is considered by most authorities as a waste product, not afterward used in the formation of the parts of the plant. Malic, acetic, citric, and formic acids, also others of less frequent occurrence than these, are found either free in cell sap or in combination with calcium or potassium. These acids occur mostly in fruits, but are also plentiful in other parts of plants.

There is another group of astringent substances which also possess a faint acid character. The most important of these are tannin, or tannic acid, gallic acid, and the astringent principle in Cinchona, Catechu, and Kino. These substances are usually found dissolved in the cell sap. Tannin is also found in little globules of solution enveloped by a delicate film of albuminous matter. It is sometimes placed in the next category, namely, among the glucosides. Nothing is positively known in reference to its function, and there are various theories regarding its origin.

15. Glucosides.

This name has been given to certain substances which by the action of unorganized ferments are broken up into glucose or some allied sugar and some other body capable of farther decomposition. The more common glucosides are Salicin, Phloridzin, and Coniferin.

16. Alkaloids.

Among the nitrogenous waste products of plants ammonia compounds are now considered the most important. Those which are not volatile at ordinary temperatures are called alkaloids; they have an alkaline reaction and are poisons. Their presence has been detected in a large number of plants, and they are especially deposited in those parts which become detached, such as bark, fruits, and seeds. There are two kinds of alkaloids, those which do and those which do not contain oxygen.

17. Crystals and Cell-sap.

Although plant cells contain a large number of crystallizable substances, the crystals contained in them are nearly all com-

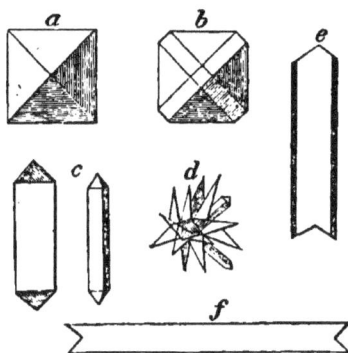

Fig. 20.

Crystals of calcium oxalate. *a–c* from leaf of *Tradescantia discolor.* *d* from cactus stem.
e and *f* from leaf of Iris. *a–c* × 300. *d–f* × 300. — (*Wiesner.*)

pounds of calcium with some acid. Crystals of calcium oxalate are by far the most common. They are found not only in cell sap but also in the walls, and are very widely distributed, nearly all plants containing them in some form. Compounds of calcium with phosphoric and carbonic acids have been found in a few instances.

Cell-sap is the name given to the watery fluid which constitutes the vacuoles. It is customary to consider it in connection with all the substances dissolved in it, which are mostly the organic compounds before described. Owing to the presence of the numerous acids it generally has an acid reaction.

PART II.

ANATOMY OF TISSUES.

CHAPTER IV. — TISSUES AND SYSTEMS.

1. True and False Tissues.

ONLY the lowest forms of plants remain unicellular during life. Ordinary plants consist of a large number of cells, differing in form and function, all of which have originated from a single one by the process of division. This is true of all multicellular plants except those originating by the so-called vegetative reproduction. In this case the new plant is detached or broken from the old and generally consists from the first of a mass of cells.

By the term tissue, in its restricted sense, is meant an aggregate of cells more or less similar in origin, shape, and function. The word is, however, often used in a less definite sense to denote any portion of the cellular substance of a plant considered as an entity.

The lowest form of what may be considered a multicellular plant growth consists of an aggregate of cells which were originally free but have united themselves for a time into a single individual. Later these separate and continue their existence as single cells. Some cell-aggregates originating in this way have a more permanent nature, their union lasting during the remainder of the life of the cells. Examples of both may be found in the low forms of algae.[1] According to the degree of

---

[1] Certain tissues forming parts of higher plants also originate by the union of cells which were at first free, for example, the endosperm of seeds. This case, however, does not fall under the head of colony-formation.

permanency of such combinations, they may be regarded either as individuals or as colonies of single cells. If we consider the colony as representing the lowest type of plant tissues, the next higher is that found most frequently among fungous growths ; this is generally called false tissue, as its mode of origin and development differs in certain respects from that composing ordinary plants. False tissue may be described as arising from a combination of branches either of a single cell, or of a number of cells, interwoven in such a way that their walls grow together, forming a compact body. Subsequent growth may occur and the number of cells may increase by new walls forming at right angles with the long diameter of the branches. The number of cells may also be increased by new branches growing and forcing their way between the old, separating the double walls at the place of their previous union. This kind of growth is found in all the higher classes of fungi, large and strong masses of tissue arising in this way. Several kinds of algae grow partly in this manner.

The term true tissue is applied to that constituting the bulk of ordinary plants. It arises from cell division, the new walls extending, in most instances, in all three directions, and at various angles with the old. Formerly it was believed that one principal characteristic of false tissue was the separation of cells once grown together, by means of new branches crowding their way between. There is now considerable evidence that parts of true tissue develop in a similar manner by a process called gliding growth, a common wall between two cells being split apart by the ingrowing of an adjacent one. If this is true, there is not so much difference in the manner of development of the two kinds of tissues. The lower grade is still marked by the new walls always forming at right angles to the long axis of growth.

Taking up now the consideration of tissues in the restricted sense of the term, it is difficult to estimate the exact number in any plant, as during its life it is subject to continued changes in which the origin and development of various tissues are involved. Again, the application of the word varies with the standpoint from which we use it. Thus we may say of a plant, in reference to its growth, that it consists of two tissues, *meristematic* and *lasting* tissue. Each of these may be subdivided in respect to other functions. If we follow the changes undergone by any plant of high organization during its development, the history and character of all kinds of tissues will be made clear; and afterward, a knowledge of their distribution in the various classes of plants is easily obtained.

2. Origin and Partial Development of the Principal Tissues and Systems.

The successive stages of development in the fertilized egg-cell of a dicotyledonous phanerogam may be described as follows. A wall is formed dividing the cell into two; others follow in rapid succession at various angles and in all three directions, making a rapid increase in the number of cells. At first all these are similar in form and all are capable of division. This continues, however, only for a short time and then localization of growth takes place; that is, a certain number of cells retain their power of division while the remainder lose this ability, being able to increase in size but not to form new cells. This gives us the first division into tissues, meristematic tissue, or dividing cells, and lasting tissue, or cells not capable of division. The meristematic tissue forms the localized centres of growth, which are at this stage two in number, namely, those of stem and root, and they occupy the opposite ends of the oval or slightly elongated body. As growth continues, the lasting tissue constituting this body is constantly added to by the new cells developed from the meristems and is thus increased in size and number of cells. Very soon after the origin of these two centres of growth, others occur which give rise to the lateral organs of the stem. In studying the subsequent changes of the plant, in order to learn the origin and development of the different tissues and systems, it will be best to select a single centre of growth, *punctum vegetationis*, and follow its history alone, leaving the other parts of the plant for special consideration.

It has been found extremely difficult to ascertain definitely the exact method of growth and cell-division at these centres; and consequently there is great uncertainty and difference of opinion regarding the meristematic tissues of the growing tips of higher plants. In fact, it is now believed by most authorities that the forms described in our text-books as illustrating the manner of growth of these plants are exceptional cases and do not represent the general or more frequent method of development. However this may be, the terms used in describing these plants have become so universal in the literature pertaining to plant anatomy, and so incorporated with it, that they must be explained and thoroughly understood from the old standpoint, before we are able to take up this study in the literature of more recent date. So far the above description answers equally well for all young embryonic plants belonging to the phanerogams. From this point onward the development of the stem apex of a dicotyledonous embryo affords the best illustrations of the tissues to be described.

At that time in the life of the young plant when localization of growth first occurs, the cells are all similar in form and size, and without definite arrangement. The first appearance of variation in shape of cells, and of any order of arrangement occurs very soon after this at the stem apex. Here the new cells derived from the little cluster of meristematic cells at or about the apex are said to separate into three groups which were named by Hanstein *Dermatogen*, *Periblem*, and *Plerome*. These three groups of cells are also meristematic in nature and they give rise directly to the lasting tissues of the stem. They are primarily derived from, and renewed by the cluster of original meristematic cells. The latter Hanstein named *initial cells* (or *Urmeristem*) while to the three groups he gave the name *primary meristems.* Each of these meristems was supposed to have its

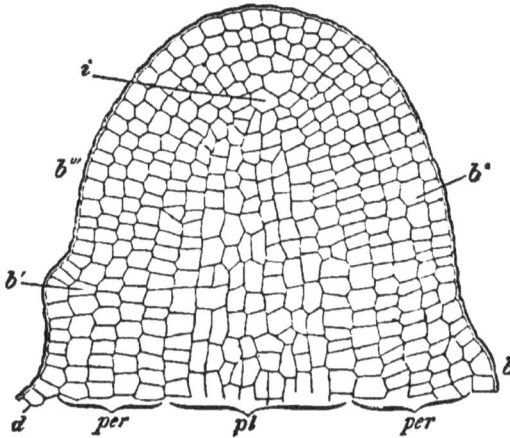

FIG. 21.

Long section through the tip of stem of *Hippuris vulgaris.* d dermatogen. per periblem. pl plerome. b b' b" beginnings of leaves. × 400. — (*Accd. to Warming.*)

own initial cell or cells, necessitating a certain order of arrangement in this little group of original meristematic cells. (Fig. 21.)

The dermatogen group consists of the external layer of meristematic cells, which remains a single layer, all subsequent divisions occurring by walls at right angles to the surface. The plerome forms a central cylinder in whose cells longitudinal divisions predominate. The periblem forms a zone, or hollow cylinder, around the plerome extending to the dermatogen, and its cells are marked by frequent and irregular transverse walls.

The plant now consists of lasting tissue and two kinds of meristems, initial and primary. All the cells are still somewhat similar in form and size, and their axes, while by no means equal

in length, are not strikingly different. In this respect they form
a single tissue, called *parenchyma*, a name applied to cells nearly
or quite isodiametric.

The next change is one which gives rise to *prosenchyma*, or
cells with one long axis, and the process may be described as
follows. A cross-section made just below the stem apex shows
clusters of cells differing from the others by their shorter radial
and tangential diameters. A long section through one of these
clusters shows that these
cells have one long diam-
eter corresponding with
the long axis of the stem.
They arise by certain
meristematic cells ceas-
ing to divide transversely
and growing faster in the
longitudinal direction, until that diameter is several
times as long as the cor-
responding diameters of
the surrounding cells.

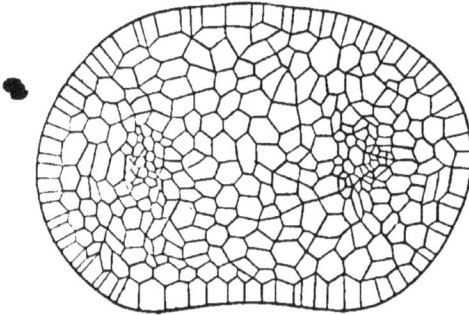

FIG. 22.

Cross-section through stem tip of *Heterocentron
diversifolium* with two cambium bundles. —
(*Accd. to Vöchting.*)

These are the cambium cells, and from them are derived the
vascular bundles. We have now the origin of the two principal
kinds of tissues as regards the shape of the cells. The cambium
cells, after reaching the necessary length, resume their meristematic nature and divide rapidly by forming new walls parallel
with their long axis of growth ; therefore cells originating from
the cambium are from the first long and narrow. Owing to this
character they are called prosenchymatic or prosenchyma, while
the cells derived from the other meristems are parenchymatic.

It is not true, however, that all cells derived from the cambium tissue remain prosenchymatic, as in certain instances such
cells form transverse walls and are thus divided into a number
of parenchymatic cells. On the other hand, there are instances

where cells derived from isodiametric meristem afterward grow long and become prosenchymatic.

If there are more than two of these cambium clusters, which is the case in the greater number of dicotyledonous stems, they are arranged in a circle about the stem axis, and are surrounded in all cases by parenchymatic lasting tissue. When they assume the function of meristem they form new walls, mostly parallel with the surface of the stem. It may also be noticed at this stage that they are more or less regularly arranged in tangential rows, and that the new cells are cut off both toward the centre and the circumference. We find these clusters of cambium developing new lasting tissue on two sides, or radially. In this way they become tangential centres of growth, isolated from each other by the intervening parenchyma. Each one of these cambium clusters with the cells derived from it forms what is known as a *vascular bundle*.

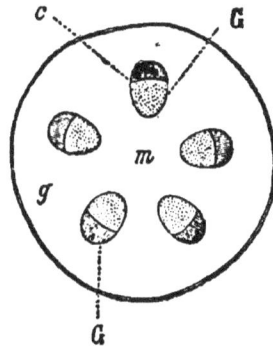

FIG. 23.

Diagrammatic sketch, cross-section of herbaceous dicotyledon. *g* ground tissue. *m* pith. *c* vascular bundles. —(*Wiesner.*)

If this growth is to continue for any length of time, it will be readily seen that some corresponding change must take place in the cells of the lasting tissue lying between the bundles. If these cells are not to be torn apart and broken by the strain, they must either grow in size to accommodate themselves to the increasing radius, or they must themselves become meristematic and so originate new cells. This leads to the discussion of a new kind of meristem, which may be called secondary, to distinguish it from that already described as primary.

It is one of the characteristics of thin-walled living cells, that they may become fully differentiated, and take on the properties of lasting tissue, and afterward assume the function of meristem cells. Such a change may take place in the tissue lying between the bundles, and it may occur in one of two ways. Either new cambium groups form between the old, and develop into what are known as interfascicular bundles, or a process takes place by which a ring of meristematic tissue is made to extend around the axis.

This ring, generally referred to as the cambium ring, is characteristic of all dicotyledonous stems, also of the woody stems of gymnosperms, and in such cases as we have described above it originates in the following manner. We have already a circle of bundles whose central cells are composed of cambium forming three or four tangential rows. The ends of these rows lie in contact with certain parenchymatic cells of lasting tissue. These lasting tissue cells now grow in length or in the direction of the long axis of the stem, until they become as long as the cambium cells ; they then assume the nature of meristem and divide by tangential walls in the same manner as do the cells of the original cambium. This change of parenchymatic lasting tissue into prosenchymatic meristem or cambium begins, as has been said, at the ends of the original cambium rows. Now suppose it to continue from group to group ; there would then arise a continuous ring of cambium around the axis of the stem.

This actually takes place, except that midway between the bundles are left a few cells which do not change their shape into long or cambium cells, but assume the function of meristem, retaining their parenchymatic form. The meristematic ring is then unbroken, but it cannot be called a continuous cambium ring. It is now easy to see how a continuous zone of new cells may be intercalated between the central pith and the outer rind, by the cambium cells giving rise to long cells like those composing the bundles, and the few isodiametric meristem cells, to the so-called medullary rays. This description of the origin of the cambium ring is based upon Naegeli's investigations, and there are known to be certain well authenticated instances where this method of development occurs. Sanio held an entirely different opinion relative to this question. He believed that the cambium ring exists first and from this ring the vascular bundles are formed. Later investigations tend to substantiate the views of both authors, and it is not at all improbable that the two methods occur in plants of different kinds.[1]

The continued development of this ring constitutes what is known as secondary growth, a subject which will be taken up later on. During the processes just described, a gradual transition from the primary meristems to the three definite systems of tissues has been completed. The word system is used to denote sets of tissues which are continuous for long distances. These are called epidermal, vascular, and ground systems, and are supposed to be derived from the three primary meristems, dermatogen, periblem, and plerome. As before intimated it has been found difficult to determine the exact limit between periblem and plerome, and it is therefore impossible in such cases to trace back the origin of the vascular and ground systems to one or the other of these two meristems. In other cases where this limit between periblem and plerome has been considered evident, it has been found that the cambium cells which give rise to the vascular system originate in part from the periblem and in part from the plerome, while the remaining meristematic cells of both classes give rise to the ground, or fundamental tissue.

The principal facts known in reference to the epidermal system are such as to offer no decided objection to the theory above explained; at the same time they are not such as to require the assumption of a special initial cell or cells for the renewal of the dermatogen. With reference to the two other systems, however, the case is quite different; here it has been found

[1] Naegeli: Ueber das Wachsthum, *etc. Beiträge zur wiss. Botanik*, 1858. Sanio: Vergleichende Untersuchungen. *Bot. Zeit.* Numbers 11–15; 47–51, 1863.

almost if not quite impossible, in any single instance, to trace
back the periblem layers to a distinct cell or set of cells, which
retain their position at the axis and act as initial cells. The
same is true regarding the plerome tissue. It is not inconceiv-
able nor impossible that the point of vegetation is differentiated
in this way, but the majority of facts now known point rather
to the opposite opinion.

Therefore at the present time the weight of evidence appears
to be against the Hanstein theory of the point of vegetation,
and it is generally believed that in all phanerogamic stems there
is a small cluster of cells irregularly arranged at the apex, which
have always retained their meristematic nature, that from this
group are derived all the subsequent cells and that just below
it the cells differentiate into the meristems from which the three
systems are developed. These cells at the apex have been named
by the Germans *Urmeristem* or original meristem. Whatever
view may be held regarding their arrangement the use of terms
is not thereby affected if it is only held in mind that the tissues
spoken of as periblem and plerome are somewhat indefinite as
to their origin and limitation.

Having now described the origin and partial development
of the principal tissues and systems, we may next begin the
study of the component elements of these systems in their
finished or mature condition.

3. Epidermal System.

This includes those cells which have the function of taking
in food materials, of throwing off or excreting waste products,
and finally of protecting the inner tissues from injurious outside influences. In reference to the last function it is divided
into primary epidermis, or that derived from the dermatogen
and continuing only for a short time, and secondary epidermis,
or periderm. As the latter occurs only in connection with

what is known as secondary growth, it will be considered under this subject. The primary epidermis is said to be simple or compound according as it consists of one or several layers of cells.

The simple epidermis covers the entire plant in the early stages of its growth, and certain organs are never provided with

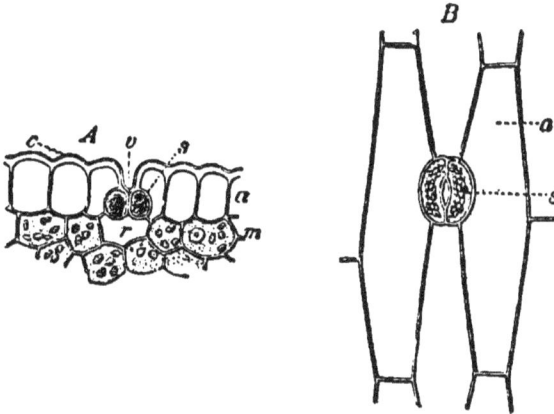

FIG. 24.

A cross-section through the under epidermis and the mesophyll *m* of leaf of *Hartwegia comosa.* *n* epidermal cells. *s* stoma with guard cells. *r* air space. *v* outer court. × 300. — (*Wiesner.*)

any other kind. Different names have been given it according to the different parts of the plant which it covers. On leaves and stems, it is called simply epidermis, on flower leaves, epithel, and on the root, epiblem. The classification may be tabulated as follows:

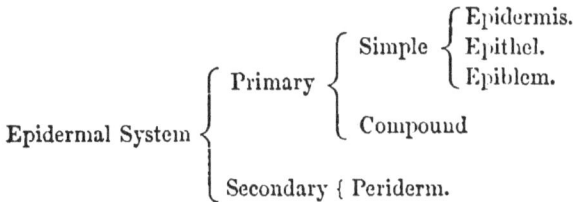

Epidermal System
- Primary
  - Simple
    - Epidermis.
    - Epithel.
    - Epiblem.
  - Compound
- Secondary { Periderm.

This consists of a single layer of cells, the larger portion of which are similar in size and shape, hexagonal with somewhat shortened radial diameter, and with no intercellular spaces. This uniformity is broken by breathing pores or stomata, and by frequent outgrowths into trichomes. The outer or surface wall of the ordinary epidermal cell is generally thicker than the remaining ones, and is also cutinized and thereby rendered nearly impervious to air or water. The walls at right angles to this are thinner and often wavy in outline. This latter characteristic is supposed to secure additional strength. In some instances all of the walls are thickened till scarcely any lumen remains, for example, in the epidermal cells of the pine leaf.

STOMATA.

These are special organs designed to allow and regulate the interchange of gases, and especially to facilitate the process of transpiration. The need for such an organ is seen from the facts that there are no intercellular spaces between the epidermal cells, and that their outer walls are of such a nature as to admit very little air or water. The stoma consists of two peculiarly shaped cells, called *guard cells*, between which is an opening leading to a large air space in the parenchyma below. The cells lining this air space connect with the guard cells and may be considered a part of the stoma, though their origin is from a different meristem than that giving rise to the epidermis. The walls of these cells are thin and consist of pure cellulose, so that both air and water may easily pass through them.

The guard cells originate early, being formed in the dermatogen layer itself. A cell divides into two unequal parts, the smaller one becoming the mother cell of the stoma.[1] The

[1] In many instances a farther division takes place so that the epidermal cells of the stoma are more than the two guard cells.

dividing wall is either straight or curved in various ways by which the shape of the mother cell is modified. This latter then divides into two equal cells, a wall forming through it which afterwards splits along its central portion, making an opening to the space below.

The guard cells at first lie in the same plane with the remaining epidermal cells, but as they develop their position may be changed so that they are either raised above or sunk below the other cells. In the latter case there is a little depres-

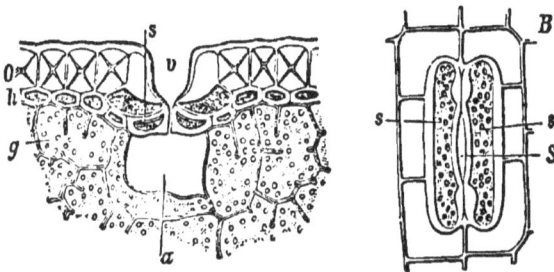

FIG. 25.

A cross-section through the upper epidermis and the neighboring tissue of the leaf of Pinus Laricio. O epidermis, h hypoderma, g green parenchyma, S guard cells with chlorophyll, a air space, r outer court of stoma. B surface appearance of the stoma, s s guard cells, S opening of the stoma. × 390. — (*Wiesner.*)

sion of the epidermis over the guard cells which is called the outer court. This arrangement has to do with the control of exchange of gases, the sunken stoma being the most common. This is farther regulated by the ability of the guard cells to open or close the aperture between them. The mechanics of this action has been studied by von Mohl in '56, by Schwendener in '81, and by others at different times, and is found to depend principally on two factors, the pressure or turgor, and the uneven thickening of wall by which a portion is left thin acting somewhat like a hinge. They are generally closed at night and open in the daytime, and with the increasing turgor of the guard cells the opening widens; with the decrease of turgor it closes. They differ from the other epidermal cells not only in shape but also in contents, being richly supplied with chlorophyll, which is seldom found in the ordinary epidermal cells. This fact is supposed to have some connection with their ability to open and close, or in other words with their condition as regards turgor.

While this is true of stomata in general, there are those whose only office is to permit the passage of water in a liquid state. Such are found on leaves of Aroideae and of the genus Tropaeoleum, and are called *water pores*. On leaves of certain plants the ordinary stomata are used for this purpose for a time before they assume the function of exchange of gases, for example, on the leaves of grasses. On the leaves of some water plants, partially developed stomata are found which appear to be entirely functionless. Certain low plants, the thalloid Hepaticae, are furnished with stomata which are very peculiar in structure and origin and have no power of motion.

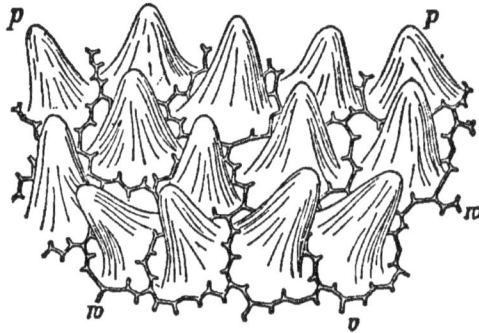

FIG. 26.

Epithel of petal of *Viola tricolor. p* papillae with striped cuticle. × 300. — (*Wiesner.*)

As a general rule, stomata are found in greatest numbers on the under surface of leaves; here they vary greatly in number, averaging from one to two hundred to the square millimeter. Sometimes they reach as high as 700, while on

the Orobanchaceae there is only one for several square millimeters.

Epithel cells are similar in form to those of the regular
epidermis, except that they are often extended externally into
small papillae. They lack other trichomes, and stomata seldom
occur. The wavy outline of the walls perpendicular to the
surface is often seen here.

Epiblem, or the covering of roots, differs but little from that
of leaves and stems. In general the cells are less apt to have
a shorter radial diameter, are very uniform in shape, and have
no stomata, except in case of large roots where the external
covering is such as to change to periderm.

TRICHOMES AND SIMILAR FORMATIONS.

Those cells or tissues which grow out from the surface of
the plant are generally known as hairs, scales, thorns, etc. According to their origin they may be classed under two heads,
those which are derived from the dermatogen, and thus belong
strictly to the epidermal system, and those coming in part from
the layers under the dermatogen, thus belonging to one or both
of the other systems as well as to the epidermal. By some
authors, the former are named trichomes, the latter outgrowths.
Those known as hairs may be either one- or several-celled ;
stems and leaves are often thickly covered with these, and they
vary greatly in form as well as in function. Some of the more
complex have the form of stem and branches, appearing when
magnified like miniature trees. Some have a single cell for a
stem on which rests a spherical head cell. *Scale* is a term given
to a short perpendicular stem terminated with a layer of tissue
parallel to the surface of the organ. All these generally have
the nature of trichomes. If they are lignified, thus becoming
hard and brittle, they are called bristles, while the term thorn
is generally applied to such as contain elements of vascular

tissue. There are a few exceptions to these rules ; for example, the thorns of the genus Rubus are trichomes, while the hairs of the stinging nettle, the gland-hairs of the rose and even the scales of some plants are really outgrowths. According to their function they may be classed : first, as secretory hairs or glands, second those which protect from too low a temperature, third, those which prevent too great evaporation. The first class includes those trichomes or outgrowths which secrete fluid or half-fluid substances of peculiar character which are generally discharged on the surface of the organ. The second, includes many short-lived hairs which die and drop off on the maturity of the organ which they pro-tect. Such are the hairs of the plane tree (Platanus) buds, supposed to fall in such quantities as to produce throat and lung diseases. Many others might be cited as belonging to this category. The third class comprises

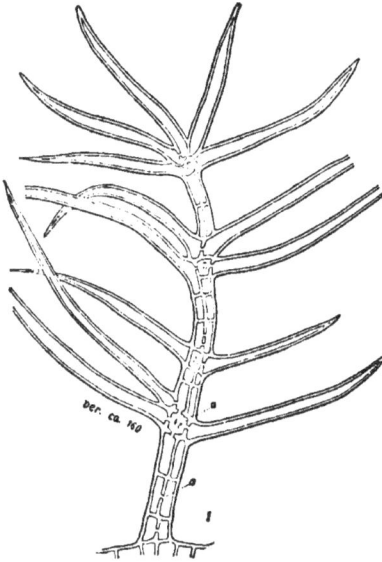

FIG. 27.

Hair from under epidermis of leaf of *Phlomis Russeliana*, partly diagrammatic. *a a* simple pores. × 160.

those hairs which grow so thickly over the surface and become so matted together that the German botanists have given this covering the name *Filz*, or felt. In this sense the term *felt* may be defined as a hair covering whose threads are so closely matted together as to leave only capillary interstices. Lastly it is believed that this felt covering has another function, namely to convey water in a liquid form to the stomata ; there are also several other arrangements of hairs supposed to promote this process, such as little tufts in the angles of the leaf-veins, etc.

The only form of trichome from the epithel is in the shape of papillae which often reach a considerable length and serve to give the beautiful velvety appearance to petals like those of the pansy.

The epiblem produces the form of trichome known as the *root hair*, whose function is so important that it has been classed as one of the essential organs of the plant. These hairs consist of a single, long, and generally unbranched cell. They develop near the tip of the rootlet and always on that portion which is just about to cease its growth in length, in this respect differing greatly from those of the epidermis proper, as these begin their development long before the organ has reached its growth. They are very numerous and extend over a zone averaging five or six millimeters long. This zone remains about the same length, as the older hairs are constantly dying and new ones are forming toward the tip. In this way the nutritive matter contained in the ground is as exhaustively covered by the growing rootlets as though each had an independent power of motion. The manner of their action in taking up this nourishment, as well as the performance of the other functions of hairs, belongs to the province of physiology.

The outer cuticularized wall of the epidermis is often covered with a thin layer of wax. This is generally absent from the guard cells of the stoma, and has never yet been found on trichomes or outgrowths. In appearance it is either that of a soft bloom as on plums, or of a crust as on the bay berry. According to the structure of the wax these coatings may be classified as follows: granular, rod-formed, crustaceous, and glazing. The granular is formed by extremely small particles which give the appearance of bloom. The rod-formed consists of parallel rods standing at right angles to the surface of the plant. This form is of rare occurrence, the best known examples being found in the culm of the sugar-cane and in the leaves of the Brazilian wax palm. The crustaceous resembles the granular except that the grains are larger and form a crust, as in the covering of the bay berry. Lastly, the glazing consists of a homogeneous coating which covers the cuticle as a continuous membrane; an example of this is found in the leaves of the *Sempervivum tectorum*, of Texas, and in the stems of Opuntia.

Organs provided with wax coatings do not become wet when put into water for a short time. As this is also true to some extent of organs whose epidermis is simply cutinized, it is concluded that all cutin contains a slight quantity of wax. Other substances of a different chemical nature from wax occur on the epidermis of some plants; such are the gold and silver coatings of certain ferns.

THE COMPOUND EPIDERMIS.

This consists of several layers of cells. They all originate, however, from the dermatogen which divides by walls running

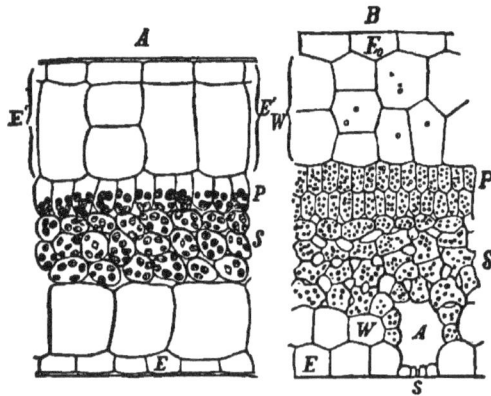

FIG. 28.

A cross-section through Begonia leaf. *E'* compound epidermis. *P* palisade tissue. *S* spongy tissue. *B* cross-section through leaf of *Tradescantia discolor*. *E* lower epidermis with stoma, *s*. *Eo* upper epidermis, both simple. *W* water tissue formed from hypoderma. × 150. — (*Accd. to Pfitzer*.)

parallel with the surface of the organ. In this way is formed a dermatogen of several layers which produces the compound

epidermis. The cells of these layers are similar to those of the
ordinary epidermis, having no intercellular spaces, containing
no chlorophyll, and supplied with stomata. The function of
the lower layers is that of water reservoirs. The best examples
are the different species of Ficus.

The secondary epidermis, or periderm, is always preceded by
the primary epidermis whose office it supplants. Since, therefore, it is a special form of secondary growth, it will be considered under that subject.

4. Vascular and Ground Systems.

It was formerly customary to consider all the tissues derived
from the cambium cells as forming a separate system, named,
from one of its important elements, the *vascular system*. More
recent anatomists consider this as only part of a system which
includes all those tissues whose axial diameter is considerably
longer than the other two. Following this method of treatment, we may use the word fibrous tissue to denote the entire
system. This includes, then, simple and compound strands or
bundles. Simple bundles are those consisting of one element
and are either collenchyma, or isolated bast cells, or sieve-tubes.

Collenchyma occurs chiefly as ribs along the corners of stems
for the purpose of support, and also as rings or zones extending
entirely around the stem. Isolated bast cells occur as sub-
epidermal ribs, in leaves and stems of certain monocotyledons.
A continuous sheath of bast cells surrounds the aërial roots of
the epiphytic orchids ; these cells originate from the dermatogen
and contain air. Isolated strands of sieve-tubes are found in
the pith of Melastomaceae, and in a large number of plants belonging to other families, monocotyledons as well as dicotyledons.
Some of the latter have them in the periphery of the pith so near
the compound bundle as to suggest connection with it. The
families Myrtaceae, Apocynaceae, Convolvulaceae and others

furnish examples of this arrangement. These bundles of sieve-tubes also occur outside the ring in the rind of the Cucurbitaceae family, and in species of Potamogeton a small bundle of this tissue is enclosed in thick-walled cells. Bast cells and sieve-tubes are both elements of the compound bundle and will be described later.

COMPOUND OR FIBRO-VASCULAR BUNDLE.

The origin of the bundles at the stem end of the plant has already been described, and it has been explained that the cambium cells may originate partly from the periblem, partly from the plerome. All compound bundles originate in a similar manner, and all agree in developing from their cambium two classes of tissue which are called *phloem* and *xylem*. In the type previously described, the bundle of the dicotyledonous stem, the phloem is developed toward the circumference, the xylem toward the centre. Each of these contains four distinct kinds of cells or elements.

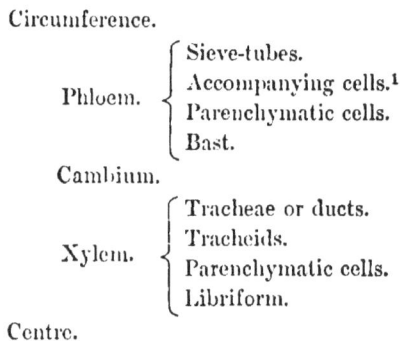

Circumference.

Phloem.
  { Sieve-tubes.
  { Accompanying cells.[1]
  { Parenchymatic cells.
  { Bast.

Cambium.

Xylem.
  { Tracheae or ducts.
  { Tracheids.
  { Parenchymatic cells.
  { Libriform.

Centre.

All these elements are not necessary to the formation of a compound bundle, as the bast and libriform may be wanting. There are different kinds of compound bundles according to the

---

[1] These cells are generally called cambiform, owing to their resemblance to cambium.

position of phloem and xylem. Before these are described the separate elements will be considered.

Beginning with the phloem, the sieve-tubes — or cribrose tissue as they are called — are its most important element. They were discovered in 1837 by Theodore Hartig, and afterwards studied by von Mohl, Naegeli, Hanstein, and others. They are found in all vascular plants but are most highly developed in the angiosperms. Here they originate, for the most part, from the cambium cells and at first are similar to these in shape and size. When fully grown, they vary much in size, especially in length. In certain climbing plants they sometimes reach the length of two millimeters. Their maximum width, which is from .02 to .08 of a millimeter, occurs in these plants.

All the walls of the sieve-tubes are thin and consist mostly of pure cellulose. The transverse walls are either horizontal, that is, at right angles to the long axis of the tube, or oblique. On these walls occur the formations known as *sieve-plates* or *fields*, which give the name to this tissue. These are described as circumscribed portions of the wall, somewhat thinner than the remaining part, and containing numerous small openings or pores. The pores are roundish or hexagonal in form, and separated from each other by narrow bands of membrane. On the horizontal cross-walls, the plates occupy nearly the whole space; and as the openings are comparatively small, the reason for the name sieve-plate is quite apparent. The oblique walls, whose surfaces are necessarily longer than those of the horizontal, usually contain several of these plates, which are oblong in form, crossing the wall so as to lie one above another, and separated by narrow bands of membrane. This is known as the *ladder* or *gridiron* arrangement.

Many sieve-plates retain throughout life the simple structure described above; others alter by assuming the condition termed by Hanstein, callous. The change consists in the thickening

of the bands of membrane between the pores, in all directions,
and it is sometimes carried so far as entirely to close them.   In

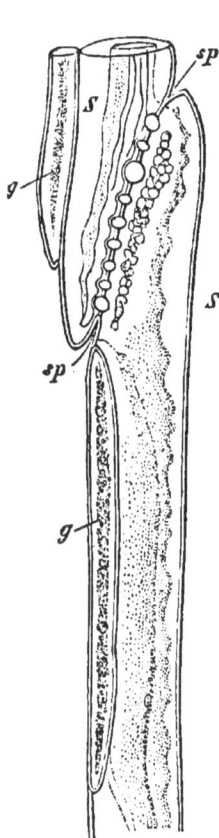

FIG. 29.

Piece of sieve-tube of *Vitis
rinifera. S S* sieve-tubes.
g g accompanying cells, *sp*
compound sieve-plates with
eight sieve areas. × 400. —
(*Accd. to Wilhelm.*)

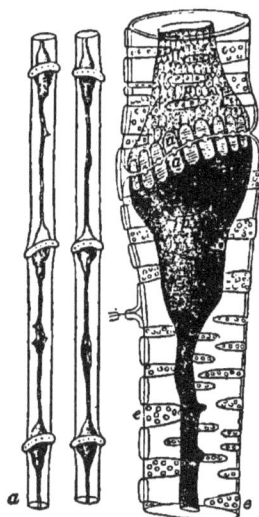

FIG. 31.

Sieve-tubes from Acer. sieve-
plates in *a* × 150, in *b* × 400
times. *a* callous substance. *p*
protoplasm. — (*Th. Hartig.*)

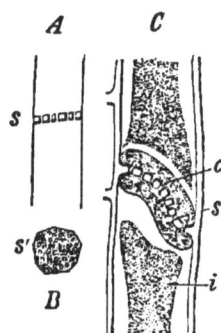

FIG. 30.

Fragments of sieve-tubes.
s sieve-plate in cross-section. *s'* seen from the surface. *c* callous substance.
× 300. — (*Wiesner.*)

some instances,
masses of this cal-
lous substance are
found,   covering
the sieve-plate on
each side, leaving no indications of the
openings or pores.   This substance is
easily dissolved, and it is believed that
by its formation the function of the sieve-
plate is arrested during the period of
winter inactivity, to be resumed again in
summer, the callous substance being dissolved away.   The contents of the tubes
are such as to support the opinion that
their function is to conduct various insoluble materials in masses from place to
place in the plant.   In their mature condition there is a thin
layer of protoplasmic substance lining the wall.   The remaining

contents enclosed by the layer are also protoplasmic in nature, but are more dense than the former. Strands of the denser material pass through the fine openings of the sieve-plates and are therefore continuous from member to member, or as is usually said, from cell to cell. This direct communication between the cells destroys their individuality; hence the term *sieve-tube* for the continued row, and *member* for each original cell.

The sieve-tube of the gymnosperms and vascular cryptogams differs but slightly from the type described above. It is apt to be prismatic rather than cylindrical. The sieve-plates occur not only at the ends of the tubes but also at the sides wherever a lateral union takes place, and they are less frequently changed to callous plates. The accompanying cells of cribose tissue are long, thin-walled, and sharp-pointed, varying but little from the shape of the cambium cells from which they arise. They follow the course of the sieve-tubes, are without pores, and always retain the character of cells, that is, are entirely enclosed by walls. For these reasons, they are supposed to be conducting cells for those nitrogenous substances which are easily dissolved.

The parenchymatic element of phloem is quite similar to all ordinary thin-walled parenchymatic tissue.

Bast cells or fibers are long and thick-walled, with pointed ends and simple pores. The name bast was formerly applied to the entire collection of tissues where such cells were found; it is now restricted to all cells of this class except those found in the xylem part of the bundle. They are sharp-pointed from the first, the transverse walls of the cambium from which they are cut off being more or less oblique to the long axis, and their subsequent growth is such as to increase this tendency. They grow in many places to several times their original length, while the surrounding tissue does not keep pace with them; it is therefore believed that the sharpened end of the bast cell is able to force its way between the lamellae of the walls which would

otherwise interrupt its growth, and by this means it retains and
enhances the peculiarity of a sharpened end.

Bast cells usually have numerous pores which are, however,
seldom circular but more often oval or oblong. Their long axis
is generally oblique to that of the cell and inclined to the left.
It is believed there is some relation between this position and
the arrangement of the micellae in the striation of the wall.
Circular pores do sometimes occur and also oblong pores with
their long axis towards the right, but these are not frequent, and
may be regarded as exceptional.

ELEMENTS OF THE XYLEM.

The ducts or tracheae are the leading element of the xylem
and correspond in this respect to the sieve-tubes of the phloem.
They were named tracheae when their chief function was supposed to be that of air passages. The name vessel was also
given them from the fact that they had lost the characteristic
mark of cellular tissue, which is that it consists of closed cells.
From this name comes the word vascular, which is still used to
characterize the entire class of elements derived from the cambium. Like all other prosenchymatic elements cut off from the
cambium, they are at first long, narrow, prismatic cells. They
afterward increase not only in length but also in the other two
dimensions and not infrequently become nearly cylindrical.
During the course of this growth the transverse walls become
either entirely absorbed, or perforated by openings of various
shapes, so that a direct communication exists between the super-
imposed cells. Another important feature in their development
is the manner of growth in thickness of the remaining walls.
This, instead of being uniform like that just described of the bast
cells, is interrupted in various ways, leaving thin places, which
allow the easy passage of fluids, while the thickened portions
furnish the necessary support. Such markings of wall have

already been described, and according to them we have spiral, ring, ladder-form, reticulated, or porous ducts. There is also another method of support used only in very large ducts,

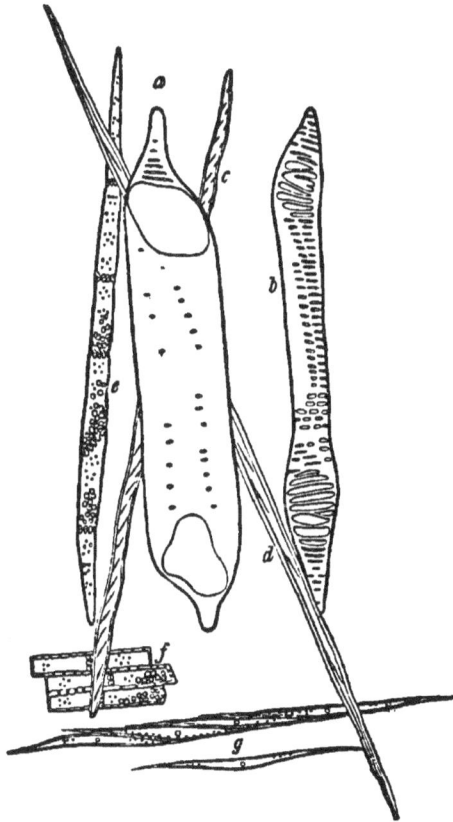

FIG. 32.

Elements of red beech in macerated condition. *a* member of a large duct ; *b* member of a small one with ladder-formed markings of cross wall. *c* tracheid with bordered pores showing only the slit-formed canals, owing to effect of maceration. *d* bast. *e* parenchyma. *f* medullary cells of narrow ray. × 300. — (*Th. Hartig.*)

namely, an irregular network of interlacing threads (or *beams* as they are called in German) which grow out into the lumen of the cell or duct.

In some cases the cross walls are completely absorbed, in others a slight edge only is left. Others, again, retain the wall, but it is either crossed by oblong openings, like a gridiron, or contains a single large central opening. In all cases the fully developed duct is no longer a living element, its protoplasmic contents being replaced by air or water, or as is more frequently the case, by alternating columns of both. In some woods the ducts are filled with resinous products, secretions of tannin and milky juices. There is also a peculiar growth found in the cavities of certain ducts to which has been given the name of tylose. It occurs where a thin-walled parenchymatic cell lies adjacent to (or borders on) the duct. At a thin place in the latter, the parenchymatic cell pushes its way into the cavity of the duct and grows out into a large sac, which may afterwards be separated from its mother cell by a wall and so entirely enclosed in the duct. It either remains in this condition, or divides into numerous cells and continues growing till large masses are formed. These tyloses have been found to be normal appearances in porous ducts of certain woods. In others, they seem to result from outward injury in such a manner as to indicate a pathological origin.

TRACHEIDS.

Tracheids are to the ducts what the accompanying cells are to the sieve-tubes of the phloem, namely assisting cells. They differ from the ducts as the accompanying cells differ from the sieve-tubes, by having no immediate connection with each other. In other words they are closed cells whose transverse walls are neither removed nor perforated by openings; their contents therefore can pass from one to the other only by osmosis or filtration. In shape and in markings of wall they resemble the ducts closely, but their average size is considerably less. It is often difficult to distinguish them from small ducts, as the only difference is that while the cross walls of the ducts are perforated, those of the tracheids are entire. In many instances, however, they resemble the libriform cells in size, length, and pointed ends, and in the tendency to thicker walls than those of the ducts. Formerly the ducts and tracheids were classed together as a single element; now the tendency is not only to consider them as distinct from each other, but to class the tracheids with the libriform tissue rather than with the ducts. Without question there are transition forms all the way between libriform cells on the one hand, and ducts on the other, and these forms pass by such insensible gradations into each other that it is difficult to say where the exact limit is.

LIBRIFORM TISSUE.

This name was first given by Sanio to that element of the xylem which corresponds to the bast of the phloem. There is

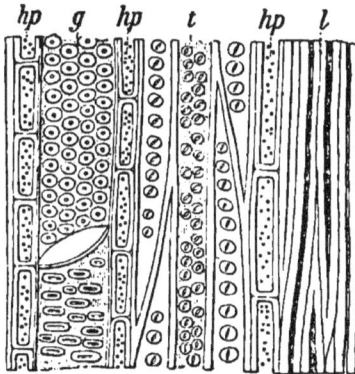

Fig. 33.

Radial long section through wood of oak. *g* ducts. *hp* wood parenchyma. *t* tracheids.
l libriform. × 300. — *Wiesner.*

little difference between bast and libriform tissue except that of location, but the degree of lignification in the walls of the latter is apt to be greater than in bast. In some libriform tissue of secondary growth, bordered pores occur; these are seldom if ever found in the corresponding phloem tissue or bast.

WOOD PARENCHYMA.

The parenchymatic tissue of xylem originates like that of phloem by the division of long cells cut off from the cambium. This may be seen from their appearance and position in a longitudinal section, where they lie in rows, parallel to the axis, each about the length of a cambium cell, and each end cell has the characteristic sharpened point of cambium tissue. It often happens that division does not take place, the cell cut off from the cambium remaining long and thin-walled, and giving rise to a tissue which can neither be classed as parenchyma or libriform, but which has received the name of intermediate tissue. Like the parenchyma of the phloem, that of the wood consists of living cells, and their position in the xylem indicates that their function is, in part, connected with the upward transfer of water.

Russow divides all vascular bundles into three classes according to the arrangement and relative position of phloem and xylem. These are,

1. Collateral—Xylem on one side, phloem on the other.

2. Concentric—Xylem in the middle, phloem around it, sometimes reversed.

3. Radial—Xylem and phloem radially arranged and alternating. These will be treated more fully in the subsequent chapters on the anatomy of the various groups of plants. The vascular bundle is in many cases separated from the surrounding tissues by a layer of cells differing from those within and without. This layer is known as the endodermis or sheath. Its contents are generally starch.

The tissues of the complete fibro-vascular bundle are tabulated by Haberlandt as follows :

| Compound or | Mestome | Leptome or Cribrose | Bast, Sieve-tubes, Accompanying cells, Cambiform | Phloem | Fibro-vascular Bundle. |
| | | Hadrome or Vascular | Ducts, Tracheids, Wood parenchyma, Libriform | Xylem | |

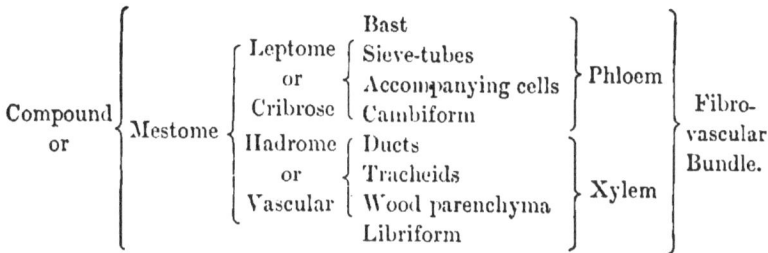

According to this table the term mestome is used for that part of the bundle which has to do chiefly with nutrition; this is subdivided into leptome or cribrose part, hadrome or vascular part. If all the elements of phloem and xylem are present, the leptome becomes phloem and the hadrome becomes xylem. Bast and libriform together are called stereome, or mechanical tissue, on the ground that their chief function is that of support.

GROUND OR FUNDAMENTAL SYSTEM.

In this system are included all the lasting cells not found in either the epidermal or vascular system. They are derived from both plerome and periblem, their position depending on the nature of the plant or organ in which they occur. For example, the arrangement of the vascular tissues differs in stem and root and there is therefore a corresponding difference in the position of the ground tissue. The cells of the latter are nearly isodiametric, and differ greatly in size. They are thin-walled, with frequent intercellular spaces, and in certain localities they often remain living and active during the lifetime of the plant.

The lasting cells of the three systems just described originate for the most part from the three primary meristems. That part of the vascular or fibrous system described as simple bundles forms an exception to this rule. Another exception is found in the phloem and xylem elements derived from those cambium cells which were not present in the original bundle. These cells, it will be remembered, originate by certain parenchymatic cells becoming (or growing) prosenchymatic and assuming the power of division. In the same way it is possible for any cell, as long as it retains its protoplasmic contents and living energy, to change from a lasting to a meristematic cell, or as it is sometimes expressed, "take on the power of wall formation" by which new cells are formed. Owing to this fact the plant has power to form organs or aggregates of cells for special functions, which cannot properly be included in any of the three systems just described. Therefore when following this classification of tissues we are obliged to add certain classes of cells as forming special organs not included in these systems. These are described in the next section.

5. Organs of Secretion and Milk Tubes.

Certain parts of the plant are known to secrete substances which are of no further use in its metabolistic processes but may be of service in other ways. These parts are known as secretory organs or glands. Glands may be defined as all sharply differentiated parts of tissue which are entirely or mostly filled with these secretions. This does not include the milk tubes, which, owing to several peculiarities, are classed by themselves. Glands are of two classes, those which excrete their contents, and those which retain them. These are again subdivided as is shown in the following table:

Glands.
1. Those excreting contents { External. Internal. }
2. Those holding contents { Tubes and tubular ducts. Intercellular channels. }

EXTERNAL GLANDS.

If wax and cutin can be considered secretions, the whole epidermal system might be included under those glands of the first class which excrete their contents externally. This, however, is hardly practicable, and it is customary to consider the glands of this system limited to certain cells and groups of cells. On some stems these groups consist of a large number of cells, as in Silene, Populus, and others. Usually the gland is com-' posed of a small cluster of cells, as on leaves of Clercodendron, where they lie just under the epidermis, which is broken by their action. Their position and the breaking of the epidermis combine to give them the appearance of a leaf parasite. The greater number of epidermal glands are hairs or outgrowths of a single cell. These seldom remain simple, but become multi-cellular. The glands of insectivorous plants are not included here, but require special treatment.

INTERNAL GLANDS.

These consist of so-called internal gland-hairs which push their way through into the intercellular spaces, and there ex-crete their contents. Such are found in leaves of Lathraceae, also in the midrib of the leaf of *Aspidium Felix-mas*. Another class of internal glands is composed of the parenchymatic cells which surround the ends of the vascular bundles where they terminate in the leaf.

Of the second class, or those holding their contents, the tubes are certain cells so filled with their secretions that nothing of the protoplasm remains or can be detected. They are in some cases elongated tubular cells, in others nearly isodiametric. The nature of the secretions varies in different plants, being gen-erally either gums, resins, mucilage, or crystals. Secretions of the latter class are apt to be found in cells called *tubular ducts* from the fact that they are situated in rows and the transverse walls are absorbed, making a true duct like those of the vascular bundle. Raphides are often found in these ducts. The inter-cellular spaces, which are utilized as organs of secretion, arise in two ways. Either the cells are separated by splits in the walls, along the line of the middle lamellae, beginning at their

intersections, or the tensions caused by growth are so strong as
to cause a break across the entire wall, subsequent absorption
removing the broken sections, till large spaces are formed
where the secretion is stored away.[1] The resin passages in the
Coniferae, Taxus excepted, and in the root and stem of *Hedera
helix*, also those containing volatile oil in roots of Umbelliferae
and some Compositae, are examples of these organs.

MILK TUBES.

These occur only in certain families and differ from other
organs of secretion in that they contain plastic material, as well

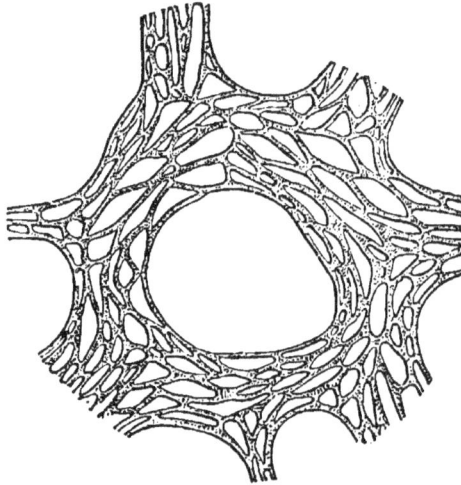

FIG. 34.

Network of milk tubes from fruit of poppy. × 150. — (*Wiesner.*)

as the ordinary secretions which are not used in the construction
of the parts of the plant. They contain no trace of protoplasm

[1] The first mode of development of intercellular spaces has been termed
schizogenic, the second lysogenic. A further distinction has been made between
those intercellular spaces which are formed when the tissues begin to differentiate — protogenic — and those formed in older tissues — hysterogenic.

or nucleus, but are filled with a milk-like fluid, generally white, seldom yellow or reddish yellow, which exudes freely when the plant is cut or broken. This liquid contains various substances, some of which are dissolved, others in the form of small grains or drops. These substances are mineral salts, sugar, gums, starch grains, albuminous matter, alkaloids, acids, wax, fats, etc. Many narcotics used in medicine are obtained from the contents of these tubes. Their principal commercial products are opium, from *Papaver somniferum*, and caoutchouc, from various plants. These tubes are richly branched and extend through a large portion of the plant, in some instances forming a continuous system, which is frequently referred to as the *laticiferous system*. The walls consist of pure cellulose, and are thin, or only slightly thickened, so that when they are

FIG. 35.

Tannin tubes in the rind of the oak. —(*Th. Hartig.*)

cut or broken the tension of the surrounding tissue readily presses out their contents, in the form of drops. According to their form and manner of development these tubes are divided into two classes, articulated and non-articulated, or multicellular and unicellular. The articulated arise from a series of elongated cells which coalesce by perforation of their cross-walls, so as to form continuous tubes. Examples of this class are found in Cichoraceae, Papaveraceae, and Papayaceae. The non-articulated arise from a single cell developing into a long branched sac whose branches grow at their apices and push their way between other tissues. Such tubes are found in the families Euphorbiaceae, Urticaceae, Apocynaceae, and others. They may be called also milk-cells in distinction from the former, which are really ducts.

Milk tubes are said to originate early in the young embryo. Schmalhausen says that in Euphorbia they start from a cell outside the plerome near the base of the cotyledon. This cell grows, branches, and extends through the entire plant, which sometimes reaches the height of a man. Others deny this and claim that the tubes originate from separate cells at the nodes of these plants. The general course which they take is such as to indicate a function similar to that of the sieve-tubes, but little is definitely known in this regard.

SUMMARY.

The foregoing description includes the principal kinds of cells and cell derivatives, as the ducts are sometimes called. The first division into tissues is that of meristem and lasting tissue. Of meristem there are three kinds : first, urmeristem, or the little cluster of cells lying at the extremity of the axis, which we call initial cells ; second, primary meristem, or the dermatogen, periblem, and plerome ; third, secondary meristem, which includes all meristem cells which were once in the condition of lasting tissue. The lasting tissues are classed in three systems, epidermal, vascular, and ground systems. In addition to these systems there are certain classes of cells forming organs for special functions, which organs are called secretion holders.

CHAPTER V. — ANATOMY OF THALLOPHYTES.

1. Fungi.

IN the low forms of thallophytes there is very little differentiation of tissues. Beginning with the one-celled fungi, which represent the lowest form of vegetable life, there is no localization of growth, the spore changing into the cell by a fairly even process of development in all its parts. The only form of tissue possible is that previously described as colony formation.

In the next higher forms, the spore in developing sends out from one or more parts of its surface tube-like or cylindrical projections, which grow at their apices only. They branch freely ; and in some species occasional cross, or transverse walls are formed, while the network of branches becomes a thin, filmy mass without solidity or regularity of form. This branched cell is known as the mycelium, or vegetative part of the plant, and the threads composing it are called hyphae.

The tissue of the fruit body originates at various points on this mycelium in the following manner. At a certain place on the mycelium a numerous and rapid growth of hyphae begins. They grow and branch freely, interlacing, and adhering wherever they come in contact, until finally, owing to the mucilaginous nature of the outer part of their walls, they grow firmly together. During this time numerous transverse walls are formed and each hypha retains its individual power of growth. In this way the pseudo-parenchymatic or false tissue of the fungi originates. If the hyphae-branches adhere only in places, instead of throughout their whole extent, and few cross walls are formed, the tissue thus resulting is loose in texture, and is called felt tissue (*tela contexta*).

Both kinds occur in some forms of the higher fungi, the felt tissue within as a sort of pith, the parenchymatic without, forming a rind.

The development of this tissue differs strikingly from that of phanerogams and higher cryptogams. The lasting tissue is not derived from any specialized meristematic cells, and the only intimation of a division of tissues into systems is that seen in the difference between the felt tissue and the firmer outer rind.

The cell walls vary in thickness, seldom show striation, and are composed of a substance slightly different from pure cellulose, inasmuch as it does not give the cellulose reaction until it has been treated with strong potash, or some other similar reagent. The protoplasmic contents show a low degree of differentiation; until quite recently no nuclei had been discovered. It is now believed that they are not infrequent. Vacuoles and a number of unorganized substances are numerous, but nothing like chlorophyll or starch grains appears. Crystals of calcium oxalate are sometimes present, but these more frequently form a crust on the outside.

2. Algae.

The lowest forms are one-celled like the low fungi, and like them they live singly or in colonies. The latter develop either by the union of cells originally separate, or by cells dividing in such a manner that while each new cell has its own separate wall, the wall of the original cell surrounds all those produced by it. These walls are of pure cellulose and in many instances show regular striations; for example, in Gleocapsa.

The multicellular forms consist of cell-rows (filaments), cell-surfaces, and cell-bodies. These vary greatly in their mode of development. Certain filamentous forms have no localized centres of growth, but increase in size by the growth of any or all the cells composing the plant. Other more complex forms show a distinct separation into meristem and lasting tissue, the urmeristem of the higher plants being represented by a single apical cell, or by a number of meristem cells. Among plants of the latter class a method of growth described as peculiar to the fungi is sometimes found. Different species of Fucus, for example, form the larger part of their tissues by cell division in the apical region, but in addition to this, long cylin-

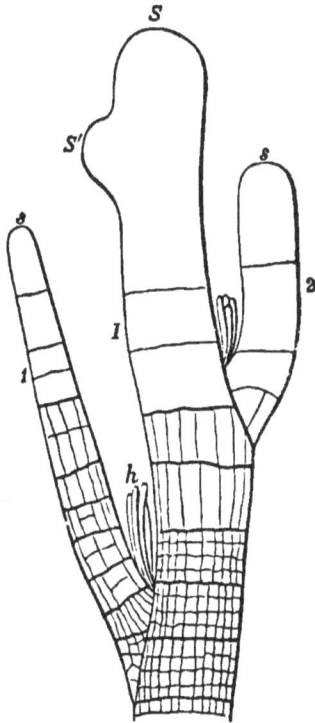

FIG. 36.

Terminal end of the alga *Stypocaulon Scoparium* Kütz. *S* apical cell. *S'* beginning of side branch. *1 2* side branches. × 84. — (*Accd. to Geyler.*)

FIG. 37.

Portion of a one-celled Alga, *Caulerpa prolifera*. (*Accd. to Schacht.*)

drical cells grow out in various parts of the plant, pushing their way between the other cells, in a manner closely resembling the growth of the fungous hyphae. A good example of growth from a single apical cell is found in the genus Stypocaulon, which grows in such a manner as to suggest the stem and

branches of the cormophytes. The apical cell of this plant is
cylindrical, terminating in a convex surface, and the axis is
increased in length by segments cut off from this cell by walls
at right angles to the main axis of growth. These segments
grow to a certain length and then rapidly divide by walls in
three directions, thus forming a cell body whose terminal portion remains as a single row of cells. Other changes occur by
which the form of the whole plant is greatly modified, but it is
not necessary to continue the description further.

The indication of division into systems is stronger in the
algae than in the fungi ; especially is this true of the outer
layers of cells, which in the higher forms closely resemble the
epidermis of phanerogams. The cell contents are quite different from those of fungous cells ; a nucleus is generally present,
and chlorophyll either in form of grains or peculiarly shaped
bands and stars. The color is often hidden by other coloring
matter, but the latter may be extracted by artificial means and
the pure chlorophyll left behind.

3. Lichens.

It is now generally admitted that the plants of this group
consist of a combination of fungi and algae, whose relation to
each other is that of parasite and host. Accepting this theory,
we cannot look for one-celled forms, nor find it possible to go
back to the beginning of growth as in case of other plants.
Experiments in cultivating lichens have partially succeeded,
but we are still ignorant of the exact processes or first stages of
growth which take place in nature. In the grown plant are
found two distinct portions, — the one of fungous, the other of
alga tissue. Generally there are two kinds of layers of fungous
tissue, corresponding to pith and rind of higher plants, then
another layer of algae, the separate elements of which were
named gonidia when first discovered. They still retain this

name, though it was given them before the true nature of their relation to the hyphae threads was known. The hyphae composing pith and rind are not at all different from those of other fungi ; their walls consist of fungous cellulose.

The algae which are found in the lichens are similar to those which exist as separate plants ; they occur either in chains or as single cells. They contain chlorophyll, and in some instances

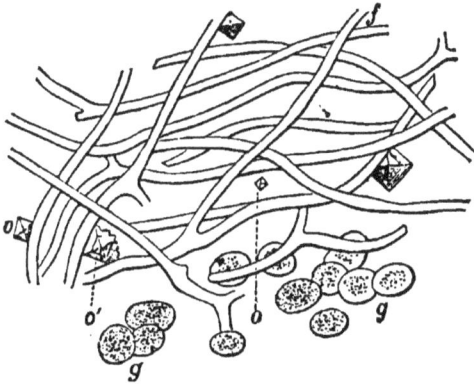

FIG. 38.

Tissue of *Rocella tinctora*, teased out with needles to show the form of the elements. *f* threads of the felt-like tissue. *o* uninjured, *o'* corroded crystals of calcium oxalate. *g* gonidia. × 300. — (*Wiesner.*)

have been freed from their fungous companions, and found to grow on as independent algae.

The most common form of lichen is a thallus, resembling somewhat that of the Hepaticae in the next higher group. To this there is no exact homologue in the real fungi. It has been suggested that the sclerotium formed by many of the higher fungi may be considered the corresponding organ. Even admitting this to be true, the mycelium form of the true fungus is wanting in the lichen.

The three foregoing classes, fungi, algae, and lichens, are generally considered true thallophytes, or plants with no distinction of stem and leaf. Among the algae, however, are

found occasional forms resembling cormophytes, and some
authorities claim that the two organs, stem and leaf, may be
clearly distinguished in these plants. It is not within the
province of this work to discuss purely morphological questions,
yet they cannot be entirely ignored, as some method of classification is necessary if the anatomical characters of plants are
to be made plain. For our purpose, it is unimportant whether
we class certain of the higher forms of algae as thallophytes or
cormophytes, for the evident transition from thallus to leafy
stem takes place in a group of plants above the algae, namely
the Hepaticae. That is, this group is the lowest in which the
distinction between stem and leaf is so evident that there is no
difference of opinion regarding it. Even here the stem does
not in all respects correspond to the perfectly radial organ of
that name in higher plants. In other words, the stem of the
leafy hepatic is dorsiventral, though in other respects it is a
true stem. Therefore simply for the sake of convenience, and
with no intention of entering into the controversy concerning
what constitutes leaf and stem, we class all plants below the
foliaceous Hepaticae as thallophytes.

In respect to the possession of stem and leaf, the group of
plants known as Hepaticae contains three forms, a simple
thallus, a thallus with rudimentary leaves, and a leafy stem.
In other respects their morphological characteristics are such as
to cause them to be classed together.

The lowest of these three forms consists of a very simple
thallus in which it is difficult to distinguish more than a single
tissue. Above this in the course of development come others
increasing gradually in complexity, but in such a way that two
distinct series are formed; the one retaining the thallus form,
but with increasing complexity of anatomical structure; the
other rising in the scale by means of outward differentiation,
such as to lead to pronounced morphological differences. The
former series culminates in the family Marchantiaceae and in-

cludes thalli with and without the rudimentary leaf, the latter in that of the Jungermanniaceae including all three forms.

One more peculiarity of this group must be mentioned before taking up the description of the thalloid hepatics separately. The difference in the mode of the development of the spore, on and after germination, has already been referred to as an intimation of the rank of the mature plant. That is, a spore destined to produce an ordinary fungous growth, germinates by sending out cylindrical tubes, which grow and branch repeatedly; walls are formed at right angles with the long diameter of these branches, the latter adhere and grow together, and in this manner the so-called false tissue is formed. On the other hand, on the germination of a spore of higher plants, a wall is formed separating the spore into two cells; each of these again divides, until a body of cells results and by a special manner of growth and formation of new cells the various tissues are produced. In certain algae there is a combination of both these methods of growth, as is illustrated in Fucus. In the Bryophytes, which includes both Hepaticae and Musci, there is still a vestige of that manner of development which culminates in the production of the false tissue of the fungi. The asexually produced spore of the Bryophytes germinates either by growing into a small plate-like structure, or by extending itself into a filamentous branched body. Both of these are called protonema, and the latter resembles the mycelium of the fungi, except that a part or the whole of it is supplied with chlorophyll.

4. Thalloid Hepaticae.

The real plant arises from this plate-like structure or protonema just described. One or more of its cells becomes meristematic and develops into an apical cell or row. The shape and manner of division of this cell or cells varies in the different families. The single cell has one outer surface wall and either two or three inner walls surrounded by the cells of the growing end. Where there are two inner walls, the new walls arise alternately, first on one, and then on the other side of the cell and parallel with the existing inner walls. Where the outer or surface wall is three-sided, the other three walls are enclosed in the growing end, making a triangular pyramid, and the new partition walls arise so that each is parallel with the third before the last, therefore at an angle of 120° with the last wall.

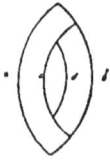

FIG. 39.

From the unequal rapidity of growth of the segments, the meristem cell or cells lie in a sinus at the growing end of the thalloid forms. The normal branching takes place in this sinus, the last segment cut off becomes a new apical cell and by its rapid growth pushes aside the original one and thus a dichotomous division is brought about. The simplest forms show no other differentiation of tissues than the possession of a midrib of several layers, while the thallus proper is composed of only one layer of cells. In the higher forms, there is an epidermis with organs corresponding to the stomata of higher plants. The thallus of the higher forms consists of several layers of cells composing different sets of tissues, which resemble those of similar function in higher plants. There is a distinct epidermis which covers both sides of the thallus, but its structure on the upper, is unlike that on the under side. Numerous epidermal cells of the under surface are extended into rhizoids, which fasten the plant to the soil and furnish it, in part, with food. On the upper surface the connection between the epidermis and the tissues below is interrupted in such a manner as to leave regular air spaces between them. A single stoma opens into each of these spaces, which are partly filled with assimi-

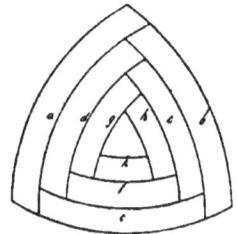

FIG. 40.

lating cells lying in rows or chains ending free in the cavity. These rows of assimilating cells spring from the tissue lying near the centre of the thallus. This tissue consists of elongated cells containing little or no chlorophyll, whose function is supposed to be that of conduction, as their long axis is parallel to that of the thallus. The rudimentary leaf occurs in these forms. It consists of a single layer of cells originating near the apex and growing out so as to cover and protect the apical cells. In this respect it has the true leaf function (that is, protection of the tender, growing cells of the apex) but in no other. It contains no chlorophyll, and its morphological rank as a leaf depends entirely on the hypothesis that the midrib of the thallus, with its accompanying wings, corresponds to the stem of the leafy plant.

From this brief description it is seen that in this class of plants there is nothing resembling the peculiar tissue which the fungous spore produces, except the short-lived protonema from which the real plant develops. Its tissues are formed in the same manner as those of the phanerogams, by division of the apical cell or cells, and subsequent growth of the new segments. Leitgeb claims to be able to follow back this process in all cases to a single apical cell, but others think that in some instances there is a row of initial cells. There is a plainly marked epidermal system, but the separation into the other two, ground and vascular systems, is not yet accomplished.

According to modern views on the relations of the different plant organs to each other, and to the homologues in the successive classes of plants, there occurs a distinct alternation of generations, beginning with the Bryophytes and extending through to the highest class of plants. In all cases the plant representing one of these generations is much smaller and simpler in structure than that of the other, and in nearly all cases this difference is so great that it is customary to consider the smaller plant only an appendage of the larger. In a work on

the elements of plant anatomy it is not considered practicable
to take up the details of the anatomical structure of the plants
of the less conspicuous generation.

Thus, in the present class, the thallus represents what is
termed the first generation, or the one bearing the sexual
organs and producing the egg-cell, which has not the ability to
germinate until after it has been fertilized. On being fertilized
it germinates at once, and begins the development of the tiny
plant known as the sporogonium, or plant of the second generation, which produces spores able to germinate without fertilization. These plants, which always remain attached to the
mother plant, have an extremely simple anatomical, as well as
morphological, structure. In respect to the latter, there is, even
in the highest forms, only an indication of a division into two
organs, stem and capsule ; while in the lower forms not even
the vestige of a stem is found, the egg-cell or spore devoting its
entire energies to the production of spores. In the next higher
class, the Musci, the corresponding plant of the second generation reaches a higher development, though here it consists only
of stem and capsule.

Chapter VI. — Anatomy of Cormophytes.

The transition from thallus to stem is said to take place in the Hepaticae of the second series, or the one formerly referred to as culminating in the Jungermanniaceae. Exactly what is meant by this expression, "transition from thallus to stem," is seldom clearly explained. It may be safe to state here that some morphologists claim that the present living forms show how such a transition may actually have taken place. The process may be described something as follows. The ordinary broad-winged thallus produces descendants with slightly narrower wings or blades; this is repeated through successive generations until finally a form is reached where the wings have entirely disappeared, the midrib alone remaining as a stem, while the organs which in the earlier generations were rudimentary are developed into real leaves.

To show this, two examples are usually taken, *Blasia pusilla* and *Fossombronia*. The former consists of a ribbon- or band-shaped thallus, bearing on its under side two rows of toothed scales, the so-called amphigastria. Leaves parallel to the long axis of the stem are also inserted along the edge of the flat thallus or stem. These leaves were formerly supposed to be portions of the thallus, but are now known to be independent organs originating from the apical cell in the same manner as the leaves of *Fossombronia*. In the latter plant the thallus is reduced to a stem, strongly flattened on the upper side, and bearing two rows of leaves inserted obliquely along its two dorsal flanks; while underneath, where the scales of Blasia occur, is found a row of club-shaped glandular hairs. These hairs are often widened at the base by subsequent cell division, so as strongly to resemble the scales or amphigastria of Blasia.

1. Foliaceous Hepaticae.

These develop in all cases from a four-sided apical cell having one outer wall and three inner, which form a triangular pyramid. In all cases except one, Haplomitrium, these plants are dorsiventral, this cell being so arranged that one of the three inner walls is parallel to the substratum. The projection of the apical cell on a plane at right angles to the substratum is therefore a triangle whose base is parallel to it. From the segments of this cell formed by walls parallel to the bases arise the amphigastria, or ventral leaves, when they are present. When not present, their place is occupied by the club-shaped glandular hairs. From the segments above arise the ordinary leaves, thus arranged in two rows. In those plants which possess the amphigastria the projection of the apical cell forms an equilateral triangle ; in those where these are wanting or imperfectly developed, the triangle is isosceles, with the shorter side for the ground line or base. In this way the character of the plant may be discovered from the apical cell alone.

The leaf consists in all cases of a single layer of cells with no distinction of blade and petiole. The shape varies, though there is a general tendency to a bipartate form. The stem is small and of simple structure. A few of the outer layers are composed of cells with somewhat thickened walls, forming a rind, while the cells within this are larger and have thinner walls. The manner of branching is varied, but the axillary branching common to higher plants very seldom occurs here.

2. Musci.

The protonema, from which the true moss plant arises, differs from that of the hepatics in several ways. It is generally a large, much-branched, filamentous body, which may give rise to several moss plants instead of one. It is usually short-lived, but may live on through the life of the plants arising from it, in which case part of it becomes rhizoid-like in character ; and in some few instances it continues to live and remains the most important part of the plant.

The moss stem. which develops from this organ, grows, in almost all cases, by means of a tetrahedral apical cell, the outer or free wall of which is parallel to the substratum. The new walls formed in this cell generally follow the law before mentioned, that is, each one is parallel to the third last before it, while in each segment cut off in this way division takes place by the formation of several new walls. The inner cells so formed go to make up the stem, while the outermost one becomes a wedge-shaped, meristem. or apical cell of a leaf. The leaf, however, develops by apical growth for a short time only, the apical cell soon ceases to divide. and intercalary growth near the base of the leaf produces the greater part of its tissue. The stem becomes a perfectly radial organ with lateral appendages or leaves. From its lower part the outer cells grow out into rhizoids which have the function of roots. The order of branching is seldom regular. In the genus Sphagnum there is one branch for every four leaves, but in the majority of cases there is no apparent law for the order or position of the branches ; they cannot be called axillary, as they not unfrequently occur between the leaves.

The tissues of the stem are more simple in character than those of the thallus of the hepatic. There is no distinct epidermal system, but on the other hand, there is in some of the higher forms a decided separation into fibrous and ground systems. The central portion of such stem consists of a single strand of thin-walled, long. cambium-like cells, with short radial and tangential diameters. In one genus, Polytrichum, there is a number of these cambium-like strands. From these cells there is a more or less gradual change into the parenchymatic. thicker-walled cells forming the several layers of the rind.

The leaf is in all cases undivided and, with the exception of the middle nerve, composed of one layer of cells. This middle nerve is lacking only in two genera, Fontinalis and Sphagnum. It consists of several layers and generally contains a bundle of long, thin-walled cells like those of the central part of the stem. These bundles run back into the stem and anastomose with its bundle or bundles in a manner quite similar to the so-called leaf-traces of higher plants.

The form just described is known as the plant of the first generation. It produces an egg-cell or oösphere, which, after fertilization, develops into a plant of the second generation. This does not separate from the mother plant, and though abundantly supplied with chlorophyll, continues to derive its support in part from the mother plant. It is called a sporo-gonium like that in the hepatics and consists of stem and capsule. The only point of interest in this case is that not unfrequently stomata occur in the epidermis of this capsule. In this respect, therefore, the plant, so simple morphologically, is further advanced in anatomical differentiation than the moss plant with stem and leaves.

3. Vascular Cryptogams (Pteridophytes).

In this class alternation of generations is quite as sharply distinguishable as in the Bryophites, although the rank of the two generations is here completely reversed. We have seen in the Bryophites, the sexual or first generation represented by the conspicuous thallus or the leafy stemmed plant. This plant arises in all cases from an asexually produced spore whose method of germination resembles somewhat that of spores of lower forms. That is, it develops first into a more or less irregular growth from which the plant springs. This plant, whether it be thallus or stem, now develops exactly as all higher plants do—that is, from localized centres of growth and with more or less sharply differentiated tissues.

Beginning with the Pteridophytes, and from this class up to the highest form, the order is reversed. The plant of the first generation, the one bearing the sexual organs, is small, inconspicuous, simple in both morphological and anatomical structure. It gives rise to the egg-cell, which, after fertilization, germinates and grows into the conspicuous and complex plant.

The Pteridophytes include several classes of plants varying widely in outward form and structure, though similar in respect to those organs which are made the basis of classification. Some of their general anatomical characters are as follows. The egg-cell begins its development by investing itself with a wall. It then divides into two cells by a horizontal wall. Further divisions take place, all the cells remaining meristematic for some time. Localization of growth occurs in such a way that in nearly all cases a root is developed from the cells originating from the lower cell of the spore, while stem and leaves are developed from those of the upper half.[1] In some instances the lower half of the egg-cell develops another organ called the *foot*, which consists of parenchymatic cells, and whose office is to connect the plant with the parent organism and to act as food conductor until the young plant is sufficiently developed to care for itself. This organ finds its counterpart in all the succeeding forms ; but as its office ceases with the independence of the new plant, it dies and disappears when no longer needed.

We have now reached that stage of plant development which includes all the organs of the highest form, namely stem, root, and leaf. The anatomy of the remaining forms is therefore limited to a comparative study of these three organs.

[1] The terms upper and lower here are used in respect to the above mentioned horizontal wall. Horizontal here refers to the position of the new plant in reference to the thallus or mother plant from which it originated, so this does not imply that the axis of the plant is perpendicular to the plane of the earth's surface.

4. Origin and Comparative Anatomy of Leaf.

Origin. — In the mosses it has already been shown that the leaf rises from a portion of each segment of the apical cell, which portion, first a single cell, afterward develops into a flat organ consisting of only one layer of cells and with no distinction of blade and petiole. In all plants higher than this the stem structure is much more complex. The presence of bundles and their connection with the leaf, and the increased complexity of the leaf itself, render its exact manner of development difficult to follow.

In all plants above the mosses, the leaf rises very near the apical region of the stem by a small number of cells near[1] the epidermis becoming meristematic, growing and pushing out the epidermis, which grows with the underlying cells so that there is no break between the epidermal covering of leaf and stem. Growth continues from the apex of the young leaf for some time. Afterward this apical growth ceases, the cells of the tip turn to lasting tissue, while intercalary growth ensues by certain cells near the base becoming meristematic and continuing the formation of new cells until the leaf is complete. From this it is seen that both epidermal and ground systems are included in the leaf. The third or vascular also forms a prominent part. At first the upper surface of the leaf grows more slowly than the under, so that its surface is concave toward the stem point, and forms a covering for it; later, it grows faster above and so straightens out. Some fern leaves grow continuously from the tip; and until the leaf is well grown the upper surface grows more slowly so that a large part is rolled completely together, thus protecting its own tip.

In the preceding description of the origin of the vascular bundles, their first appearance near the tip of the stem was

---

[1] In some few cases the leaf rises from the dermatogen layer itself, as in *Elodea Canadensis.*

described. We shall now have to return to the same figure to illustrate the manner of connection of leaf and stem by means of these bundles, which form the ground-work of support for the leaf as well as the stem. In the example before given, the bundles are described as forming in the stem and belonging to it. In this respect there are two classes of bundles, those which are formed in the stem and always remain there, growing in acropetalous order, and those which grow for some distance in the stem, and then curve outward into a leaf, so that they belong both to leaf and stem. The former are called, in German, *stammeigene*, or those peculiar to the stem, and they may always remain entirely disconnected with any leaf bundle, or the latter may extend from the leaf base downward and attach itself to the stem bundle. Those bundles which belong to both stem and leaf are called leaf-trace bundles, and in respect to their place of origin and subsequent development they also form two classes. Those of one class are said to originate in the stem and bend out into the leaf; of the other, it is said that they originate at the base of the leaf and run in both directions, outward into the leaf blade, and downward into the stem until they reach the stem bundle and unite with it.

In point of fact, the leaf originates in the stem and its base is in the stem when the bundle arises, but it is customary to describe the course of the bundles in this way. The stem, therefore, may have leaf-trace bundles only, or stem bundles with separate leaf-trace bundles, or stem bundles with leaf-trace bundles attached. The leaf-trace bundle does not always enter the stem entire, but often divides into two or more strands which enter the stem at a node and run down through one or more internodes. Their subsequent course varies according to the varying species.

Comparative Anatomy of the Leaf. — The simplest form of leaf has already been described, namely, a single plate of cells in the Bryophytes. In the leaves of all plants above the Bryo-

phytes, there are found three distinct systems, epidermal, vascular, and mesophyll, the latter corresponding to the ground
system of the stem. As has been seen from its method of development the leaf does not possess any urmeristem cells, therefore never ends in a bud.

In the Pteridophytes a sharply distinct epidermis covers both
sides of the leaf. The inner portion consists of roundish or
parenchymatic, chlorophyll-holding cells, with frequent and
large air-spaces between, and vascular bundles disposed so as to
give support and conduct material to the chlorophyll-holding
mesophyll cells.

In the phanerogams the structure is still more complex, the
mesophyll cells being usually separated into two distinct classes
of tissues each consisting of one or more layers. These tissues

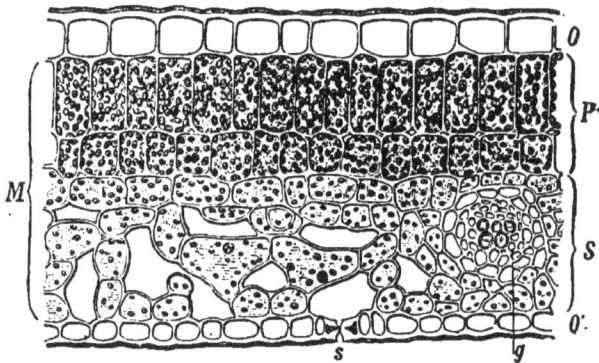

FIG. 41.

Cross-section through leaf of *Cyclamen Europaeum.* *O* over, *O'* under epidermis. *S* stoma.
M mesophyll. *P* palisade tissue. *S* spongy tissue with vascular bundle, *g.* × 300. —
(*Wiesner.*)

are called palisade and spongy tissue. The former occurs
usually immediately below the upper epidermis and consists of
one or more layers of oblong cells, so situated that their long
diameter is at right angles to the surface of the leaf. Intercellular spaces are very seldom found between the cells of this

tissue and it also connects with the epidermal system without air-spaces, forming in this respect a very sharp contrast to the spongy tissue. This (or the latter tissue) lies below the palisade cells and in connection with the lower epidermis. Between the palisade and spongy tissues run the bundles of the vascular system, which form the framework of the leaf. The spongy tissue is like the mesophyll of the cryptogamic leaf, except that its cells are more irregular in shape. They often send out arms in different directions between which are the large air-spaces peculiar to this tissue, and continuous with those of the numerous stomata. There is also frequently a hypoderma or layer of cells immediately under the epidermis, whose office is to support the epidermal cells. In certain fleshy plants this hypoderma consists of several layers and is supposed to act as a water reservoir. Examples of this are leaves of Ficus and Tradescantia.

Nervature. — The ridges or nerves evident on nearly all leaves are formed by the strands of bast or collenchymatic tissue which occur usually in clusters on the under and upper sides of the bundles. The nervature therefore coincides with the course of the bundles. These are branched in all the phanerogams, though for systematic purposes some are described as unbranched. Even in the parallel nerves of the monocotyledons minute branches stretch out from the large ones, anastomosing with them and with each other.

Through the petiole the bundles generally extend straight toward the blade, and are arranged either in an arc open toward the upper surface of the leaf, or in a closed ring, or sometimes with no order but scattered irregularly over a cross-section of the petiole. In the blade they spread out in different directions branching and anastomosing, their course being especially varied in dicotyledonous leaves. The two parts of the bundles are so arranged that the phloem is toward the upper, the xylem toward the under surface of the leaf. The bundles either end free along the edges, or, as is more frequently the case, they unite, forming a continuous nerve near the edge.

The general characters of the phanerogamic leaf may be stated as follows: the epidermis of the upper surface is nearly, if not quite, without stomata; that of the under surface has numerous stomata connecting with large air-spaces. The cells between the two epidermal layers are divided into palisade and spongy tissues; the former lies next the upper surface, and is richly provided with chlorophyll; the latter lies below, its cells containing less chlorophyll, and the bundles of the vascular system run through its upper portion.

Sometimes there is present a hypoderma and frequently sclerenchymatic[1] cells of different shape are scattered irregularly through the central portion. Thus the leaf is a dorsiventral as well as a bilateral organ. A striking difference in structure is shown in those leaves which for some reason, such as twisting of the petiole, infolding of the blade or other departure from the ordinary position, develop the palisade tissue on what is morphologically the under side, and the spongy tissue above, thus exactly reversing the normal order of arrangement.

In leaves which are more or less cylindrical in shape, or which differ greatly from the flattened organs so far described, the tissues are inclined to a concentric arrangement, similar to that found in the stem. This is true in the so-called pine needles and other Conifer leaves. These are mostly designed to live for several years instead of a single season, and therefore, as might be expected, the anatomical structure differs from that described in the typical leaf. One example may be taken from the genus Abies. A cross-section of this leaf shows peculiarly thickened epidermal cells, and next these, a closely attached hypodermal layer, broken only by the stomata. These latter occur at somewhat regular intervals around the whole circumference. Next come numerous layers of green parenchymatic or mesophyll cells thickly interspersed with resin

[1] See note on page 125.

channels, and then a bundle sheath surrounding the two bundles which are farther enclosed by colorless parenchymatic cells. These two bundles extend the whole length of the leaf and are entirely unbranched. The position of phloem and xylem corresponds to that found in the ordinary flattened leaf.

In the scales of the winter buds, which are metamorphosed leaves, we find a similarity in general structure to that of the real leaf. The under epidermis has more strongly thickened outer walls. Stomata occur only on this surface, and, what is peculiar to this kind of leaf, the primary epidermis is sometimes replaced by periderm, a tissue which will be described later on as belonging to secondary growth. For example, *Aesculus Hippocastanum*.

Calyx leaves have stomata only on the under side. The mesophyll is often divided into palisade and spongy tissue. The bundles have mestome strands only, with no bast.

Corolla leaves have a uniform mesophyll (not separated into palisade and spongy tissue). The ducts of the bundles are mostly spiral, these together with thin-walled fibrous cells making up nearly the entire bundle. The epithel has very few stomata, and its cells are often papillose. The coloring matter occurs either in the epithel or parenchymatic cells, or in both, and is usually dissolved in the cell sap. Starch grains and calcium oxalate crystals are of frequent occurrence, and also ethereal oil, the last contained in glands, oil channels, or suspended in little drops in cell sap of epithel or parenchyma or both.

Cotyledons, in seeds containing no albumen, consist of richly developed parenchymatic cells, epidermis without stomata, and bundles distributed throughout. The contents of the parenchyma are oil, aleurone grains, starch, etc.

When the function of the ordinary deciduous leaf is completed, its separation from the stem is brought about in the following manner. A plate of cells at its base, where it joins

the stem, becomes meristematic. They grow, divide, and are
filled with cell sap. This evaporates in such a manner as to
cause a tension, which the new and weak walls are unable to
bear ; they break and in this way a complete separation occurs.

5. Comparative Anatomy of the Stem.

Up to this point in the description of the stem, the entire
development has been spoken of as coming from the meristem
near the apex ; this meristem lies entirely above the protuber-
ances described as the beginnings of leaves. These organs play
an important part in the history of the growth of the stem, and
it is owing to their connection with and influence over it, that
its manner of growth is so different from that of the root. To
make this clear, certain morphological relations must here be
explained. It is usual to consider the methods of leaf arrangement as reducible to two, opposite and alternate ; the first,
where two leaves appear to start from the same height on the
stem and opposite each other ; the second, where the leaves all
start from different heights on the stem, always with some regular order of arrangement as regards their distance from each
other measured on the circumference of the stem.

According to the usual method of growth, the beginnings
of the formation of the leaf organs follow each other so rapidly
in the apical region that quite a number are formed before the
oldest succeed in growing out to any considerable size, and
while the longitudinal distance from one leaf to another is
almost infinitesimally small. In case of winter buds the older
leaves grow out as scales and extend entirely over the young
stem tip, with its season's growth of embryonic leaves and internodes ready for development in the early spring time. The tip
itself with its tender meristem tissue is protected by some of
the inside leaves which entirely cover it. When the season for
renewed activity begins, the actual or visible extension in the

length of the stem takes place by the rapid growth and division of the very small cells between those points on the stem where the lateral organs occur. In this way, the young leaves, while still growing outward with their own independent meristem tissue, are stretched apart by this intercalary growth of stem.

The place on the stem where the leaf separates entirely from it, fixes the position of what is called a node, that is, the circumference of the stem marked by a line drawn around it through the base of the leaf, and at right angles to the long axis of the stem. That portion of the stem between two nodes is called an internode. In some of the monocotyledonous stems the base of the leaf extends around this whole circumference and the nodes are marked externally by prominent ridges. In stems with opposite leaves, the leaves of each pair stand so nearly at the same height, that it is convenient practically to consider them as standing on opposite sides of the same node, the internodes in this case being the separate lengths of stem between the pairs of leaves.

Thus the stem, while constantly being added to, and while owing its actual increase in number of cells chiefly to the meristems at its apex, owes its real extension in length to the subsequent growth of the new cells so derived. The number of cells is also added to during the growth of the internodes by cell division, since it is one of the characteristics of lasting tissue that it may at any time take on the character of meristem, provided only that it still retain its protoplasmic contents.

The apex of the moss stem and that of nearly all of the vascular cryptogams consists of a relatively large and actively dividing apical cell. From this division results a building tissue which usually shows no division into meristems corresponding to those said to occur near the apex of the phanerogamic stem. When, however, there is an intimation of such a division into meristems, both in the mosses and vascular cryptogams, the outer is supposed to correspond to the dermatogen and periblem, the inner to the plerome of the phanerogamic stem.

As the simple structure of the moss stem has already been explained, we may pass at once to the consideration of the stem of the vascular cryptogams or Pteridophytes. A few of these have a cluster of initial cells instead of a simple one at the stem apex, and the plants possessing such a structure may be considered a link between the vascular cryptogams and phanerogams. The stem contains compound vascular bundles which are generally concentric, and in all cases closed. The ground system is sharply distinct from the epidermal layer, which is supplied with stomata. In all the young plants of this class the bundle system of the stem may be considered a sympodium built up of leaf-trace bundles, a single one of which runs into the stem from each leaf. In most cases the first bundle starts in the foot of the embryonic plant, where it ends free. From the outer extremity it bends out into the first leaf or cotyledon, and at the point where it bends outward another new bundle starts, which runs along in the stem a short distance and then bends outward into the second leaf. In this way arises an axillary bundle which may be considered as made up entirely of leaf-traces.

This is the simplest possible arrangement of the stem and leaf bundles, and we have seen it already intimated in the stem and leaf of Polytrichum, where the place of the real bundle is supplied by long fibrous cells. In a number of Pteridophytes this simple structure remains throughout the entire life of the plant. In others, for example those of the Lycopodiaceae and some of the Selaginellaceae, the manner of development of the young plant is such that the axillary bundle of the grown stem is rather to be considered as belonging solely to the stem, while its corners are formed by the leaf-traces.

Another large class of Pteridophytes begin the development of their bundle system as above described, continuing this

manner of growth till the fifth or sixth leaf is formed ; then
with a sudden enlargement of the diameter of the stem, another
method of arrangement ensues which may be described as a
widening out of the axillary bundle into a hollow cylinder.
so that there is a central pith and an outer rind.   Since the
parenchymatic cells of the pith act partly as conducting cells,
and partly as reserve cells for the food material of the plant,
it would be very disadvantageous to have them entirely isolated
from the parenchymatic cells of the rind by a
continuous zone of vascular bundles.   A connection is therefore secured between the pith
and rind by means of what is known as the leaf
opening, namely, an aperture in the bundle
tissues just below the insertion of each leaf.
In some plants this is only a small opening just
below the nodes or places where the leaf separates from the stem.   To this class belong, for
the most part, plants whose stems are small
creeping rhizomes with alternate two-ranked
leaves.

Fig. 42.

Course of vascular
bundles in stem of
Blechnum boreale.
a branches of the
vascular bundle
running down-
ward toward the
leaves, slightly
magnified. —
(*Accd. to Unger.*)

The greater number of ferns with upright
stems, leaves in many rows, and short internodes,
differ from the above type in having larger
openings separated from each other by relatively
smaller or thinner bands of vascular tissue. so
that the hollow cylinder formed by the bundle
tissue appears to consist of a network whose
meshes are the so-called leaf openings.   As this type is of
frequent occurrence, a description of a single stem may be
given here in order to make the manner of construction clear.

The young plant of *Aspidium Felix-mas* begins its growth
with leaves arranged with an angular divergence of one-third,
and with their separate bundles united sympodially into an
axillary stem bundle.   Having reached the fifth or sixth leaf,

the stem increases in diameter, the one-third leaf position changes to three-eighths, and the net-formed tube begins at the insertion of the last leaf of the one-third arrangement.

This may be illustrated by a diagrammatic sketch. In Fig. 43 let the shaded bands represent the vascular tissue, and the white or unshaded portions the meshes; the points of insertion of the leaves will then be at the places indicated by

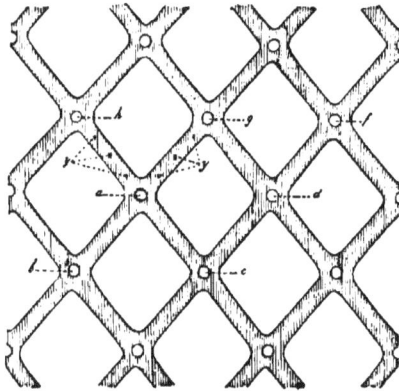

FIG. 43.

Explained in text.

a, b, c, etc. The leaf whose insertion occurs at a has a single bundle bending out at this point, which is formed by the union of two others originating at the points b and c, or the points where the bundles of the leaves b and c bend outward into them, their position in reference to point a being always that of the two next older leaves on either side. The repetition of this forms a net with rhomboid meshes. Each leaf receives also four to six bundles from along the sides of the network. In large and well developed plants, each leaf receives therefore as many as seven bundles, one at the lower angle of the mesh and six above, three on each side, as indicated at y. In the next year the stem becomes much thicker, the leaf position going over to five-thirteenths, where it usually remains.

The bundle system in the foregoing example is considered as made up of leaf-traces, and this is also true of that of many other ferns having this cylindrical network. In many kinds of fern-stems, however, it is impossible to ascertain the exact relation between leaf and stem bundles; while in others it is decided that the network, otherwise similar to that described, consists mainly of stem bundles, the leaf-traces being attached to these. There is still a third method of arrangement found in stems of certain ferns with many rows of leaves. Such plants show on a cross-section of stem several concentric bundle rings, each of which forms a net-like cylinder as before described. On a long section these several hollow cylinders have the appearance of conical mantles widening toward the apex of the stem. There are numerous places of connection between these mantles, caused by the anastomosing of the leaf-trace bundles, making the actual course of development extremely difficult to discern.[1]

In several instances the bundles of fern-stems are collateral instead of concentric ; for example, Osmunda. A secondary epidermis or periderm is never developed on the fern stem, but frequently its place is supplied by a well developed hypoderma.

BRANCHING OF STEM.

The anatomy of the branches, or secondary axes as they are sometimes called, is similar to that of the primary stem. The exact place of origin and the first stages of development of these side axes is often very difficult to determine. In the Pteridophytes there is much uncertainty respecting the manner of branching in the various classes. Some few facts of general application have been ascertained, but before stating these, a brief description of the morphological relations of stem and branches may be inserted.

There are two principal systems of branching, the monopodial and the dichotomous. In the monopodial system, the main axis continues to prolong itself, its meristems remaining active, while the side branches rise from a new cluster of initial cells, near the apex in acropetalous order. All these new branches are of the same order or rank, in respect to the

[1] For a more detailed account see De Bary's Comparative Anatomy.

primary stem, and since this is called a foot or podium we have the name monopodium. These side axes may in their turn branch, producing stems of the third order, and these again till the system is completed.

In the dichotomous system, the main axis discontinues its growth, and two new branches start on the opposite sides of the stem apex at the same level. Each of these develops into a new axis which may branch again in a similar manner.

There are several variations of these two methods; for example, the sympodial method, which may be considered a special form of the monopodial. In this case, when the side branch starts to grow, the main axis ceases to develop. The side branch grows on and often takes an upward direction until it, in turn, branches; it then ceases to grow, and its side branch grows on, prolonging the apparently simple axis, which is however, composed of lateral axes of different orders; hence the name sympodium, an apparent axis, for this system. When the different branches composing this axis take an upward direction, the outward appearance of the stem so produced is so similar to the real monopodium that it is extremely difficult to distinguish its true character. There are various other ways by which the normal order of both monopodial and dichotomous methods of branching is changed, but these three forms are most frequent. Among the ferns the branching is either dichotomous, monopodial, or a combination of both. In the Salviniaceae and Marsiliaceae it is plainly monopodial, while in many ferns it has not been determined whether the method of branching is monopodial or sympodial. In these doubtful cases the secondary branches cannot be said to be axillary, though their position may sustain a certain fixed relation to that of the leaf. They arise from the stem either at the side of the place of leaf insertion or above or below it, their bundle system being united to that of the stem in various ways. There are also other examples of dichotomous branching where the entire bundle system may be considered double, for example, the Lycopods and Selaginellas, although in these plants, the forking is not always equal.

There are also adventive branches or buds which may occur at any time and almost at any place along the stem. The bundle system of such branches unites itself to the nearest portions of that of the stem. To these few facts, it may be added that in general, the connection between the bundle systems of branch and main axis is such as is best adapted to the various functions of the tissues concerned.

PHANEROGAMIC STEMS.

According to their morphological development the gymnosperms should be placed next to the vascular cryptogams. The anatomical structure of their stems, however, corresponds to that of the dicotyledons, and in a comparative study of the anatomy of the vegetative organs of plants, the stems of gymnosperms and dicotyledons may be considered together, while the structure of the monocotyledonous stem is on an entirely different plan and of such a nature as to rank it next to that of the Pteridophytes.

As regards the development of the respective centres of growth of the young embryo, it has been stated that the apex of the stem of vascular cryptogams consists generally of a single cell, but that in a few cases, occurring chiefly in the family Lycopodiaceae, a cluster of cells has been found to hold such a position in reference to the stem axis that the cells composing it may be called apical. That is, they lie about the axis in such a manner that each may constantly divide and give off new cells without losing its own position and rank. For a long time it was supposed that the apex of some Conifer stems was of this nature, and efforts were made to prove that even here, examples might be found of growth from a single apical cell. If this could be proved it would furnish another evidence of the possibility of genetic connection between the vascular cryptogams and the gymnosperms. Some botanists carried this so far as to claim that the meristem cells at the apex of all phanerogamic stems might be traced back to a single apical cell. This, however, has never been accomplished in a single instance, and the generally accepted theory of the arrangement of the meristems is that known as Hanstein's, to which reference has already been made. In order to show more clearly what is meant by this theory and what are the real facts known at the present day, two examples may be described, one taken from the vascular cryptogams, the other from the phanerogams.

The manner of growth from a single apical cell is at first very simple. It divides, making two cells, one called a segment which is added to the cells of the axis, and another which retains the form, character, and position of the original apical cell and again divides, cutting off a new segment cell, which also enters into the structure of the axis. The Equisetum stem furnishes an example of this class. Its apex consists of a three-sided pyramidal cell, which cuts off segments by walls parallel to the three sides of the pyramid. In the tissues of the growing region of the stem apex which are derived from these segments, there may be distinguished two classes of walls with reference to their direction. These are periclinal, or walls parallel with the surface of the stem; and anticlinal, or those at right angles to the surface and so to the periclinal walls. The outer layer of cells, which later forms the epidermal system, does not extend over the apex of the stem above the origin of the leaves. On the contrary periclinal walls are formed in the outer layer of that portion of the apex which is above the leaves. Within the outer layer there is no separation of the cells into different layers. It can only be said that from the cells lying near the central portion come the pith cells, while from those lying near the periphery come the epidermal layer and the zones containing the bundles. (Fig. 44.)

The manner of growth at the apical regions of phanerogamic stems is quite different. That of the stem of *Hippuris vulgaris*, a common water plant, may be taken as an example. This is described by Hanstein about as follows. A longitudinal section through the axis of the stem tip shows such an arrangement of cells as to preclude the possibility of growth from a single apical cell. The periclinal curves run very regularly around the tip so

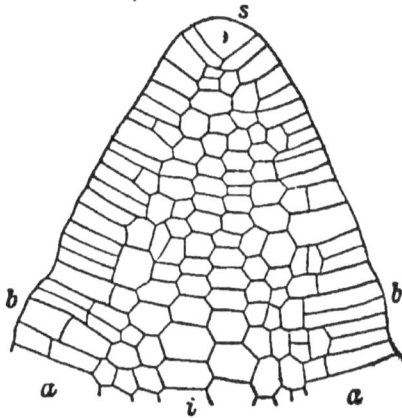

FIG. 44.

Point of stem of Equisetum.   *s* apical cell.   *a* outer, *i* inner meristem.   *b* beginnings of leaves.   × 300. — (*Wiesner.*)

as to give the point of vegetation the appearance of being made up of separate layers of cells. The anticlinal curves may be clearly seen though they are not so regular in their course. The entire point of vegetation is here covered with a single layer of cells, which with few exceptions divide always by anticlinal, never by periclinal walls. If this layer of cells be followed down toward the older stem portion, it is found to pass directly into the layer which forms the epidermal system. This layer of actively dividing cells, covering the entire tip of the stem, Hanstein named dermatogen, and claimed that at its summit is a cell or group of cells which give rise to and perpetuate it. He

named this the initial cell or cells according as there were one
or several. (Fig. 45.)

That part of the stem tip lying under the outer layer, is
composed of a meristematic tissue from whose outer portion are
produced the cells originating the bundles, while the inner
layers, or group of cells, give rise to the tissues forming the
central part of the stem. These inner layers Hanstein named

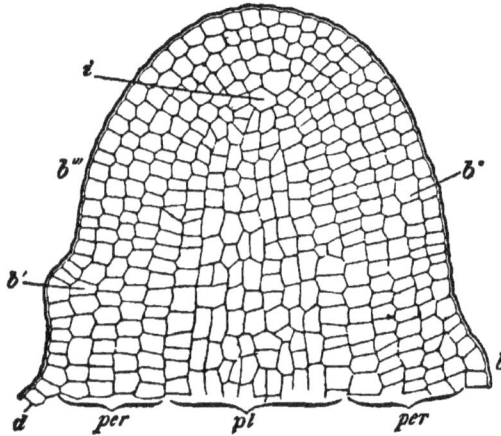

FIG. 45.

Long section through the tip of stem of *Hippuris vulgaris*. *d* dermatogen. *per* periblem.
pl plerome. *b b' b''* beginnings of leaves. × 400. — (*Accd. to Warming*.)

plerome. They were also supposed to end toward the stem tip
in a group of initial cells corresponding to those of the dermatogen layer. So with the layers of cells lying between the
plerome and the dermatogen, which he named periblem ; these
too were supposed to terminate in a group of initial cells.

This interpretation of the order and development of the
various meristems of the point of vegetation of both stem and
root was believed to apply to a large number of phanerogams.
The examination of a great number of such plants, however,
has failed to corroborate this opinion. In some cases where

there is a distinct dermatogen there is no apparent separation of periblem and plerome ; and in other instances where such separation is plainly indicated by direction of walls and arrangement of cells, it is quite impossible to trace back the meristems to distinct initial groups. Although much time and labor have been spent on the study of the apical regions of phanerogams in order to ascertain their exact manner of development, no definite, satisfactory results have as yet been obtained. It is now believed by many that the urmeristem of such stems, from the Conifers upward, consists of a group of cells with no definite order of arrangement and no regular method of division. The terms introduced by Hanstein are those in common use when referring to these regions, and there is no danger of their misapplication if these facts are held in mind. With this explanation we may now pass on to the consideration of the different kinds of stems.

MONOCOTYLEDONOUS STEM.

A cross-section through such a stem discloses at once the peculiar structure common to this class ; that is, the bundles appear scattered singly and without order over nearly the whole cross-section. They are usually more numerous toward the periphery, and in many instances there is left a central pith quite free from bundles. This either remains during the life of the plant, or becomes destroyed so as to leave a hollow stem. The bundles are always closed and generally collateral, with a tendency to the concentric arrangement, as both phloem and xylem are frequently surrounded by a bast cylinder. The primary epidermis is well developed, and in certain cases is replaced by periderm. The ground system is mostly parenchymatic and sometimes appears to be divided into rind and pith cells, as there is often a small zone near the circumference comparatively free from bundles. Although there are several varieties of stems of this class in reference to the arrangement

of vascular tissue, the main features of its method of growth
are as follows.

Leaf-trace bundles occur in every stem, and in many there
are also stem bundles. When the latter occur they run near
the periphery and nearly parallel with the long axis. The
peculiar course taken by the leaf-traces in the stem is the

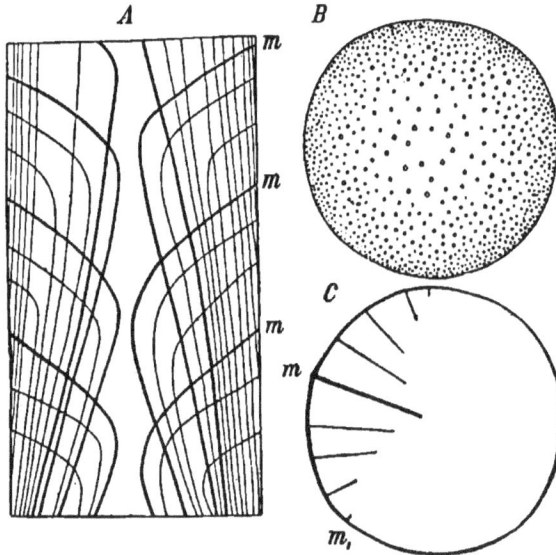

FIG. 46.

Diagrammatic representation of the course of the vascular bundles in a monocotyledonous
stem. *A* radial long section. *B* cross-section. *C* horizontal projection of the leaf-
traces. *m* medial, *m'* marginal nerve. — (*Wiesner*.)

cause of the scattered arrangement of bundles on the cross-
section. There are usually a large number of bundles from
each leaf which enter the stem separately, and take a radially
oblique course; in some large plants the number reaches
several hundred. The leaf basis extends entirely around the
stem circumference, and from this basis each bundle continuing
downward enters the stem bending strongly toward the centre,

and then back toward the circumference, its course becoming more nearly parallel with the stem axis as it approaches the circumference. After running down through several internodes, it either unites with the leaf-traces of older leaves, or ends free. The depth to which a leaf-trace penetrates the stem differs for the bundles of the same leaf. The midvein reaches farthest into the stem axis, those of the margin extend the least distance, while those remaining take an intermediate position. This bent course of the leaf-trace is generally accomplished in a radial direction, that is, the leaf-trace lies in a plane parellel with, and cutting the long axis of the stem, but traversing this in an oblique direction ; while in some exceptional cases it does not lie in a single plane, but makes its way toward the axis by bending tangentially also.

The stem bundles are often simple strands of bast or collenchyma which usually appear near the circumference in the form of epidermal ribs. In some stems similar strands also occur near the centre. There are many variations from the type described, an important one being that of the Commelynaceae family which is similar to the dicotyledons in the arrangement of its vascular tissue. There is a circle of stem bundles lying near the periphery, at some distance from the leaf-trace bundles which are near the centre. The stems of Dioscorea and Tamus are the nearest to the dicotyledons, as here the leaf-traces are arranged in a single circle about the circumference. In their course, however, the bundles penetrate radially into the stem, extending to unequal depths, while those of the dicotyledons never approach the stem centre.

In certain long-lived monocotyledonous stems, growth in thickness is accomplished by means of a secondary meristem forming near the periphery of the stem. It differs from the so-called secondary growth of the dicotyledons, because the additional thickness is caused by the repeated intercalation of new bundles and their subsequent growth, while in the dicotyledonous stem the original bundles take part in the formation of the new growth.

Regarding the development of the bundles in the embryo, present investigations show that in some cases the bundle of the cotyledon runs down into the axillary one of the root ; or if the cotyledon has more than one bundle, these unite at its node and from there run down united into the root bundle. The bundles of the embryonic leaves immediately following the cotyledon take the course described for the mature plant, so far as the smaller number of bundles and internodes will permit.

STEM OF DICOTYLEDONS AND GYMNOSPERMS.

There are more variations in the character of the stem in the dicotyledons and gymnosperms than in any other class. These variations consist largely in the different courses taken by the bundles and in the manner in which they anastomose with each other. With these exceptions the greater number of these stems may be referred to a certain type considered as the normal one.

The mature stem is covered with an epidermis which is supplied with stomata. This epidermis is generally replaced by a second one named periderm. The ground tissue consists mostly of thin-walled parenchyma, with occasional strings of collenchymatic cells.

The bundle system of most dicotyledons, Conifers and Gnetaceae — Welwitschia excepted — may be described as follows. The bundles are collateral and open. All the primary ones are leaf-traces. They enter the stem bent at the nodes and run downward radially in a straight course about midway between the centre and circumference of the stem. The stem therefore consists

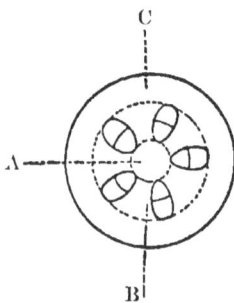

FIG. 47.

a central cylinder or pith. b hollow cylinder enclosing bundles. c hollow cylinder composed of primary rind.

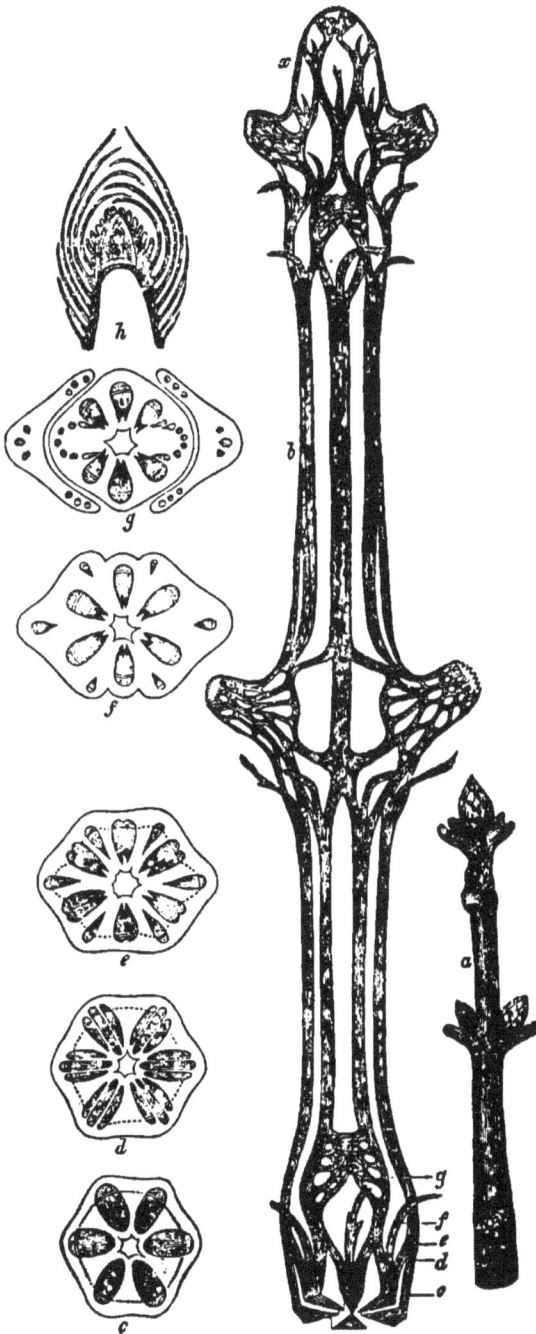

FIG. 48.

a tip of stem of *Atragene Alpina;* *b* skeleton of the same obtained by maceration. *b* an internode consisting of six bundles as seen in the cross-section, *c*. *h* the young bud with the leaf sheaths. *g* shows the cross-section through stem and leaf petiole at the point marked *g* on the skeleton figure. The leaf-traces of each leaf are separated into nine bundles. At *f* three simple leaf-traces are enclosed in the tissue of the stem, but still remain outside the circle of bundles. At *e* and *d* they enter into this circle and are fused with it.

of three distinct parts, as may be seen on a cross-section ;
a central cylinder or pith ; a zone surrounding this, which may
be considered a hollow cylinder, of which the bundles form the
chief part ; and outside of this another hollow cylinder consisting of the primary rind. (Fig. 47.)

All leaf-traces which consist of a single strand, and nearly

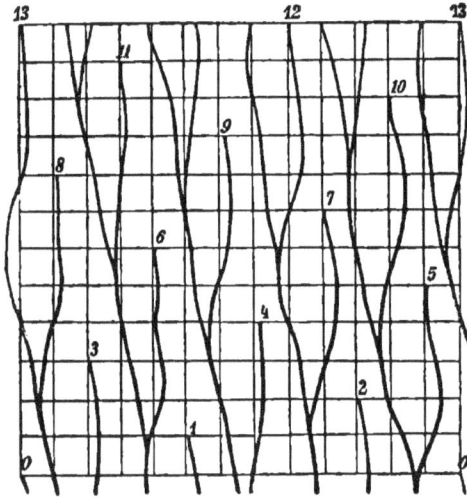

FIG. 49.   FIG. 50.

Diagrammatic sketch of the tangentially oblique course of the   Plan of the course of the
bundles in the young stem of *Iberis amara*. The bundles   bundles in stem of *Juni-
which form the surface of a cylinder appear here spread   per nana*. *s* leaf-trace
out in a plane. *0*, *1*, *2*, etc., places where the leaf-traces   bundles. *k* bundle of
enter the stem. — (*After Nägeli*.)   bud. — (*After Geyler*.)

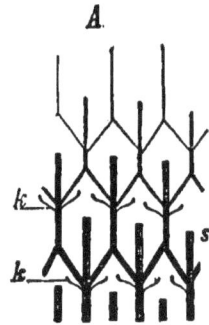

all those with divided strands, run downward through more than
one internode and then anastomose laterally with those of older
leaves. That is, they are so inserted between the bundles already in the circle that they do not break its course, and in this
lateral union of the bundles similar parts always come in contact
with each other, the xylem of the new with the xylem of the
old, and so on. (See Fig. 48.) The union of the new bundles
with the old occurs either at the nodes or very near them, and

is either one-sided forming a sympodium, or the two arms join so as to make a network. (Figs. 49 and 50.)

We may now compare the two types. the mono- and the dicotyledonous in respect to the chief differences in the course of their bundles. In the latter, the course of the bundle is radially straight, but tangentially oblique ; while in the mono-cotyledon it is just the opposite, it is radially oblique and tangentially straight. To illustrate this, successive cross-sections of the dicotyledonous stem present the same appearance, a circle divided into three parts, pith, bundle-zone, and rind-zone. If we wish to represent the course of the several bundles, we must consider a whole or a part of the hollow cylinder which they compose as taken out of the stem and flattened on a plane surface. We shall then have such a figure as 49 or 50, — the lines in Fig. 49 representing the course of the leaf traces with a single strand, and in Fig. 50 that of those with divided strands.

If we wish to show the course of the bundles in the mono-cotyledonous stem, we must cut a radial section from the centre to the circumference, and that will contain the bundle from its entrance into the stem to its point of intersection with the older bundle, or till it ends free. Of course, reference is made here only to those stems which may be considered as following the normal type. There are many plants which vary from this. For example, the tangentially oblique course of the leaf-traces of dicotyledons is not common to all. In many stems they run nearly parallel with the long axis, bending only at the nodes. This is usually the case in stems with opposite leaves. It may also be mentioned here that when their course is tangentially oblique, they seldom continue long bending in the same direction so as to form a spiral, but more often they alternate from right to left, thus taking an S-formed course. In a very few dicotyledonous stems the bundle system is so arranged that the course of the stem bundles is radially oblique, like that of the monocotyledons.

A few general exceptions from the so-called normal type may be mentioned here. In certain stems are found what is known as the bicollateral bundle. For example, *Tecoma radicans* has two rings of cambium cells, the outer one developing normally, phloem outward and xylem inward. The inner one joins the xylem on its outer limit and develops only phloem, and that toward the centre. Another similar form occurs in Cucurbita, Nerium, and other stems, where the regular cambium of the circle of bundles at first develops phloem in both directions; and after a time it ceases to produce phloem toward the centre, and begins to develop xylem in that direction. Such double rings of phloem are found in quite a number of plants.

In certain other stems the normal circle of bundles is supplemented by other circles, which originate either at the point of vegetation, as does the normal one, or from a secondary cambium which rises in the primary rind. In the first case, the rings consist generally of leaf-trace bundles, and they may lie either within the normal circle or without it; that is, in the rind or in the pith. When in the pith, there may be one or more circles of the bundles, or as in the cotyledons they may lie scattered irregularly on the cross-section of the pith. In stems of Umbelliferae, Orobanchaceae, Begoniaceae, there are stem bundles in the pith. Finally, in many swamp and water plants belonging to dicotyledons there is in the stem a central vascular bundle, which either arises from the union of several leaf-traces, or it may be considered a single stem bundle united to those of the leaves at the nodes. Examples of this class are Hippuris, Myriophyllum, etc. Such central bundles are also found in several monocotyledonous stems, as in Corallorrhiza and Najas.

Stems of the Coniferae and Gnetaceae are similar in general structure to those of the dicotyledons. There is, however, a histological difference in the Conifer stems which does not appear in the primary growth; that is, the xylem of the first year's growth contains all the elements common to it, while in all the succeeding years' growths no ducts are formed, their place being taken by tracheids.

As might be inferred from the character of the open bundle, which is normal only to the dicotyledonous type, these plants are designed to prolong the growth of their main axis in all three directions, while plants whose stems have closed bundles increase in diameter only until the bundle system has reached its full development. After this time the stem increases in diameter only in a few exceptional forms. This continued increase in the diameter of dicotyledons and gymnosperms is of such a character that it can best be treated in a separate chapter. The process by which it is accomplished is generally termed secondary growth, and this takes place in root as well as stem. It may therefore be omitted until the description of the general structure of the root has been given.

The normal method of branching in all stems above the vascular cryptogams may be described as axillary; that is, the branch originates in the axil of the leaf. Those branches occurring elsewhere on the stem are termed adventitious, there being no law governing their place of appearance. It is not possible here to enter into a description of the various methods of growth of these secondary organs, but it may be said in general of those that are normal that their bundle systems originate in the periblem of the stem tip, and like the bundles of the leaves usually from the outer layers of the meristem. For this reason both sets of organs are called exogenous in referring to their place of origin. The mode of attachment of their bundles to those of leaf and stem varies greatly, but it takes place in such a manner that not only is there a direct continuity of the main-stem and side-branch bundles, but in the greater number of instances the pith of the main stem immediately joins that of the side branch. For a more complete discussion of this subject the reader is referred to De Bary's Comparative Anatomy and also many more recent monographs.

6. Comparative Anatomy of the Root.

We come now to the consideration of an organ about whose general structure and function there is no possible ambiguity. Regarding its morphological value, however, there is room for a difference of opinion. It has been customary to speak of all organs whose function is to hold the ordinary plant to the soil or the parasitic plant to its host and which conduct food from these to the plant, as either true roots or root-like organs. To the latter class belong all such organs of plants below the vascular cryptogams. True roots occur then first in the vascular cryptogams and are common to all plants from this group upward. There are, however, a few even of the highest class of phanerogams which have no root. The most prominent distinction made by botanists between the root-like organ (rhizoid) and the true root is that the true root is always tipped with a root-cap. The presence or absence of this cap determines the rank of the organ, though functionally both classes are the same. If we inquire after the reason of this apparently arbitrary test, it will be seen that there is some ground for this distinction with which the root-cap itself has no connection, except that it is always on roots having a certain origin. It is therefore only a sign of their method of origin.

If we go back to the simpler forms of plants to find the first traces of root-like organs, we may perhaps begin with the haustoria which grow on the mycelium of certain fungi. A little higher up in the scale, we reach what may be taken as the normal type of this organ in all plants below vascular cryptogams, namely: the rhizoids of the lichens, Hepaticae, and Musci.

The transition from thallus to cormophyte is said to begin in the group Hepaticae and to reach its full completion in the next higher group, the Musci. If we compare a thalloid plant of the former class with a cormophyte of the latter, we find that the rhizoids of the two plants are of equal rank, as well in respect to their place and manner of origin, as in their structure.

To illustrate this we compare the thallus of Marchantia with the stem of any of the mosses. The former is a flattened, bilateral, dorsiventral body, whose point of vegetation is so situated with respect to the direction of the main growth of the plant as to insure this taking place in a horizontal direction and from one apex only. The rhizoids spring from the side of the thallus turned toward the earth or substratum. In the moss the main body of the plant is a radial organ composed of a complex of cells growing mostly from one extremity, while the rhizoids arise from a region directly opposite to the growing apex, fastening the plant to, and feeding it from the soil.

The moss is a perfectly organized cormophyte. It does not, however, originate directly from the development of a single cell, but arises as a secondary stage of an asexually produced spore or cell. The egg-cell which it produces, on being fertilized, gives rise to a plant of extremely low morphological character; we must therefore pass to the next class of plants to reach a root of different character, or a true root.

In this class, namely the vascular cryptogams, the relation of the two generations changes and the plant that represents the species, that is the plant with stem and leaf, arises not from the asexually produced spore nor from any of its phases of growth, but from the fertilized egg-cell itself. Here then is the origin of the real or true root. It is an embryonic development of the fertilized egg-cell in such a way that one end of the main axis of the plant becomes the organ functioning as the root. In the early or embryonic form of such a plant, its organs may be described as two, axis and leaf; but the axis develops in such a way that it forms two organs, stem and root. Side organs of similar character are given off from both of these axes, those from the root being always roots, and those from the stem either roots or stems. In comparison with stem and leaf, the root is characterized by great uniformity of anatomical structure, but this uniformity is found only in the lasting tissues. As regards the meristems at the point of vegetation there is greater complexity than at the corresponding point in the stem. The three primary meristems described as belonging to the stem are said to occur here and also a fourth is added in some instances, while in their manner of growth there are said to be, at least, six types.

This complexity in the growing tip of the root arises from the peculiar nature and function of the root-cap. As the root never produces leaves, it lacks their protection over the soft and tender growing tissues of its apex, and this protection is supplied by the cap, which is a lasting tissue surrounding the meristems of the root tip. It is made up of parenchymatic, polyhedral cells. The inner rows of these cells, or those lying near the meristem, contain no intercellular spaces, but as new cells are added and the older rows pushed outward, their cells begin to separate at the corners and the tissue becomes less firm. This outer portion of the root-cap is subject to the continued friction of the soil through which it is pushed, and in this way its cells are gradually separated and thrown off from the rest of the root, their places being supplied by the cells lying next to them. This process is constantly going on, thus necessitating a continual regeneration of the cap cells. So the meristem forming this organ is constantly active, although the cap itself is not extended in length as is the root proper. By this manner of growth the meristematic cells of the root tip are protected from friction and also from too rapid transpiration. In aërial roots, the latter is the only danger to which they are exposed; consequently their anatomical structure is somewhat different.

The other lasting tissues of the root are grouped into three systems corresponding to those of the stem, epidermal, vascular, and ground systems. Like those of the stem, these also are supposed to arise from three primary meristems, dermatogen,

periblem, and plerome. Besides these, in certain instances there
is an independent, separate meristem named calyptrogen which
gives rise to the root-cap. Thus there are in the most complex
roots, four distinct primary meristems. Others lack the calyp-

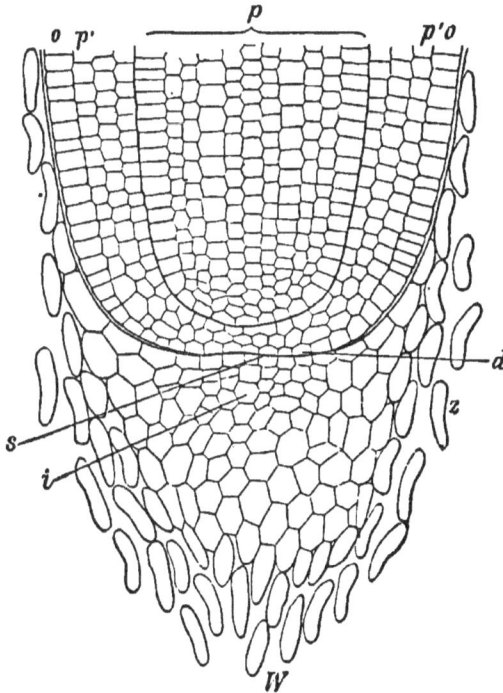

FIG. 51.

Longitudinal section through root tip of *Zea Mais*. *W* root-cap; *i* younger part. *z* cells
loosening from cap. *s* point of vegetation. *o* epidermis. *p'* periblem. *p* plerom. *d*
dermatogen bounded on the under side by the calyptrogen. × 150. — (*Wiesner*.)

trogen, the cap tissue there being derived from what is termed
the reciprocal division of some of the other meristems. That
is, the meristematic cells give off new cells in two directions,
some outward, forming the cap, others inward which are added
to the long axis of the root. The farther consideration of this
subject belongs to a more advanced course in anatomy.

Regarding the urmeristem, or the cell or cells from which the primary meristems are derived, it has been found here as in the stem tip, that the roots of most vascular cryptogams have a single apical or urmeristem cell. In some roots of this class, however, there has been found an urmeristem consisting of a few definitely arranged initial cells. There is the same uncertainty respecting the structure of the point of vegetation of phanerogamic roots, as there is in respect to that of the stems of this class of plants. Here, as in the stem, it has been found difficult to ascertain the exact manner of growth of the cells which give rise to the lasting tissue and what has been said of the Hanstein theory applies equally well to root and stem.

The epidermal system consists of the so-called epiblem cells. These differ in various ways from the cells covering organs exposed to light and currents of air. They are more uniformly isodiametric, contain fewer stomata and have only one kind of trichome, which is known as the *root hair*. From a physiological standpoint these root hairs hold a very important position in the economy of the plant, as they are the organs by which it obtains food from the soil. They are simple in structure, consisting of a single epiblem cell prolonged into a thin-walled hair. They begin to develop a short distance from the root tip and extend entirely around the root, covering a portion several centimeters in length. As the root grows on from the tip, the hairs are constantly renewed, the oldest dying away in about the same proportion that the new hairs are added.

Beginning now with the root of the vascular cryptogams, it may be added that in these the epiblem is not replaced by a periderm, although a well developed hypoderma frequently takes the place of the epiblem layer, which dies and disappears. The ground system consists generally of parenchymatic starch-holding cells, which are bounded outside by the hypodermal layer, and inside by the endodermis which surrounds the central vascular bundle or bundles. The vascular system is composed of a central cylinder made up of a single radial bundle, or, as it may be considered, a collection of several. The radial bundle is common to all roots with very few exceptions, and it occurs

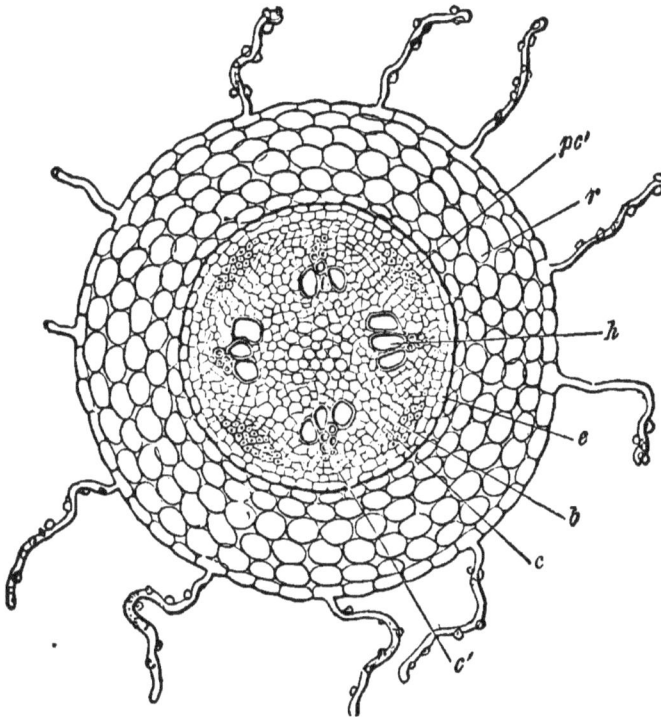

FIG. 52.

Cross-section through a young root after the bundles are started. Under the epidermis with the root hairs is the rind parenchyma, *r*, which is bounded within by the endodermis, *e*, which later forms the cork layer. Inside the endodermis is the pericambium. *pc'*. The bundles are arranged so that the wood alternates with the bast. The cambium ring, *c*, which later develops the secondary growth, lies inside the bast at *c* and outside the wood at *c'*. — (*Th. Hartig.*)

in only a very small number of stems. In it the xylem and phloem groups are radially arranged, that is, they lie next each other but on different radii. The simplest interpretation of this arrangement would be to consider the number of bundles equal to the number of xylem clusters, as for each of these there is a corresponding gróup of phloem. It may, however, be quite as logical to regard the whole cylinder as a single central bundle, and those who prefer this explanation give to it the names monarch, diarch, triarch, polyarch, etc., according as there are one, two, three or more xylem groups. In the root of most vascular cryptogams there are two or more such groups. (Fig. 52.)

The root of monocotyledons differs but little from that of the cryptogams in respect to its vascular system. In some plants there is a well developed pith occupying the central portion of the cylinder, the whole being surrounded by the endodermis or sheath. Some few monocotyledons, for example Pandanus and a few palms, have roots in which the bundles are separate. The ground and epidermal systems are also similar to those of the cryptogams except that the hypodermal sheath of the cryptogamic root is generally parenchymatic, while in the monocotyledons it is more frequently a regularly developed endodermis.[1]

In the early stage of their development the roots of dicotyledons resemble those described above in all important features. In case of many annuals this construction remains during the entire life of the plant, and of these it may be said, the structure of their root is the same as that of the monocotyledons. It is, however, very different with other annuals, with perennials, and all woody or long-lived plants. In these there is a later development corresponding to that before referred to as secondary growth of stem, but it differs from this in that it begins much earlier in the life of the plant. The complicated structure of the root tip and its various methods of growth render the developmental history difficult to follow in all its

[1] The name endodermis has been given to the cylinder or sheath surrounding a bundle or collection of bundles. This sheath consists of a single layer of cells derived from the ground system ; they are longer in the axial than in the other two directions, usually contain starch and their walls are frequently suberized.

details, but a general description of the beginning of the bundle cylinder and its subsequent development may be given as follows.

The origin of the root bundle differs in several respects from that of the stem. This may be illustrated by a cross-section near the root tip where there are four xylem plates. At their first appearance they consist of four cambium strands similar in character to those described in the stem, that is, they are composed of long, small, sharp-pointed cells arranged in groups each of which presents an oval circumference on the cross-section. These four ovals may meet at the centre, in which case there is no pith formed. The first appearance of xylem is produced by certain of these cells turning into lasting tissue of the nature of xylem. This development or change proceeds at first toward the centre, and continues until the entire central portion is occupied by xylem, a large portion of which consists of ducts. The extreme outer portion of the cambium string remains meristematic and after the xylem at the centre is completed or during this time, it continues to develop or cut off new cells toward the centre, which are also in time turned to xylem tissue. Between the plates of xylem the cells do not at first grow into cambium cells to give rise to phloem plates, but remain parenchymatic. They do, however, become meristematic, cutting off new cells so that the section which they compose keeps pace with the radial increase in the xylem plates. The first appearance of phloem is caused by some of the cells of the outer portion of these sections changing gradually into phloem elements, chiefly sieve tubes, but these are often accompanied by bast cells. (Fig. 53.)

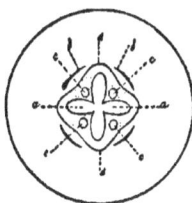

FIG. 53.

a a represent the cambium bundles from which are derived the first xylem plates. *b b* the phloem elements of the sections between. *c c* the new xylem plates caused by secondary growth. *d* the cambium ring.

This gives an approximate idea of the manner of origin of a radial bundle. In roots where a pith occurs the process is similar, but the four cambium clusters do not meet at the centre, thus leaving a cylinder of ground cells which form the pith. This also illustrates the so-called primary growth in all cases whether of annuals or perennials. At this stage the meristematic cells lie at the outer part of the xylem plates and within the phloem elements. Now, when secondary growth is about to take place it is first indicated by the formation of four new xylem plates which are developed from the meristem inside the phloem. These plates like the old, are distinguished by large ducts. A continuous cambium ring is now formed by the meristem of each phloem-plate gradually extending itself till it meets that of the xylem plates. This ring develops new cells both centripetally and centrifugally, those toward the centre forming xylem elements, those toward the circumference, phloem. From this time on the process is like the secondary growth of the stem, the various elements exactly corresponding, except that in the root there are no primary medullary rays. When this growth is continued from year to year, the radial increase in the central portion of the root demands a corresponding one in the tissues of the circumference or rind. This is brought about by the formation of a periderm followed by bark. As this growth differs in no essential manner from that of the stem, known as secondary growth in thickness, farther description of it will be given in connection with this subject. The roots of the gymnosperms are similar in all essential particulars to those of the dicotyledons.

In very many plants the primary root dies early, and its place is supplied by side roots. These rise endogenously, that is, they arise from a little cluster of cells inside the endodermis of the central cylinder. These become meristematic, grow and push their way through between the other tissues and finally break through the epidermal layer. The bundles of the side root are connected with those of the mother axis in various ways ; hence arise a number of different types in this respect.

Roots may also branch dichotomously in the same way that the stem does, for example, Selaginella, Isoetes, and other cryptogam roots.

There are one or two peculiarities in the manner of the axial growth of the root which may be mentioned here. As the root develops normally under the soil, the stretching or growth of the cells axially is confined to a much shorter distance than in the stem. Thus growth in the length of the stem occurs in a number of internodes simultaneously, some of which are at a considerable distance from the stem apex, while in the root this stretching of the cells in direction of the long axis is probably confined to the space of a centimeter, at most, back from the apex and in many instances probably to a few millimeters. It is by this sudden lengthening of the end of the root that the tip is driven into the ground. Another peculiarity is the shortening or shrinking of the roots of certain plants, particularly those which are herbaceous. This gives a shrunken appearance to the rind and a winding course to the bundles. The result in case of herbaceous plants is to draw the stem part closer to the ground while in woody plants it seems to fasten the whole more securely to the soil.

Note. — In order to simplify as much as possible, all unnecessary terms have been omitted in the foregoing chapters on tissues and systems. The word sclerenchyma is used by some authors as the name of a tissue including all thick-walled cells. Others limit its meaning to parenchymatic, thick-walled cells, like the stone cells of fruits and rinds. We have preferred to use it only as an adjective to describe any cell with a hard, thick wall. Hypoderma also is used here, not as the name of any special tissue, but of the layer immediately under the epidermis when the character of the cells composing this layer is changed from the normal type.

CHAPTER VII. — SECONDARY GROWTH IN THICKNESS OF STEMS AND ROOTS.

THE process known as secondary growth in thickness, is common to the stems and roots of most dicotyledons and gymnosperms. It occurs also in some of the larger stems of the monocotyledons, but their normal stem is not planned to allow for yearly increase in radial diameter, and the stems which do admit of such increase may be considered as having changed from the mono- to the dicotyledonous type. The origin of the bundles in the dicotyledonous stem and their manner of distribution have already been given while describing the elements of the vascular system. We may now return to the same figure used there for illustration, namely, a cross-section of an ordinary dicotyledonous stem. The bundles as a rule are collateral, open, and all leaf-traces; their course in the stem is such that if the entire section containing them, were taken out from the stem, or separated from other parts of it, this section would be a hollow cylinder. There would remain two other parts, a central solid cylinder or pith, and an 'outer hollow cylinder or rind. Growth in thickness of such a stem can be accomplished through the activity of the cambium cells lying in the cylinder containing the bundles, but only on one condition, that is, that the ground cells lying between the bundles also increase in radial diameter either by the growth of the cells already present, or by some of them becoming meristematic and so forming new cells. In stems designed to live for some time, the latter method is followed by the formation of the so-called cambium ring.[1]

---

[1] As there is so little uniformity among authors in the use of anatomical terms, for convenience, we have used the word cambium to denote only the prosenchymatic meristem, in order to distinguish it from the parenchymatic.

This ring is formed in different ways, according to the peculiarity of the different plants where it occurs. In general it may be said to be formed either by the intercalation of new bundles, or by the formation of interfascicular cambium. The new bundles are either leaf-traces or they belong to the stem alone. New leaf-trace bundles are introduced when the arrangement of leaves on the stem changes so that the upper part bears more leaves than the lower. Both leaf-trace and stem bundles are so introduced that their respective parts exactly correspond with those of the older ones. The cambium of the old and new bundles does not, however, form an unbroken circle as there are, in all cases, a few of the ground cells left between the bundles, which cells become parenchymatic meristem, thus completing the circle of meristem cells; though the cambium is interrupted by as many of the parenchymatic meristem groups as there are bundles in the circle.

These groups give rise to the primary medullary rays as they cut off cells in both directions, outward and inward, keeping pace with the growth of the cambium cells. On the cross-section they appear very similar to those of the cambium, and it is only on the long section that their real character is made evident. The primary medullary rays thus formed extend through the bundle cylinder from rind to pith, and serve as conductors of cell contents. Their

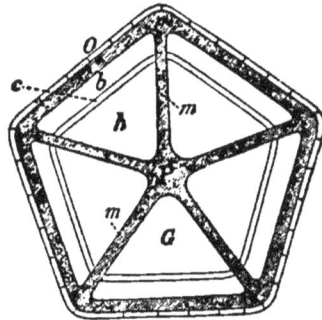

FIG. 54.

Diagrammatic cross-section through a young dicotyledonous stem. *O* epidermis. *p* primary rind. *m* primary medullary ray. *b c h* one of the five vascular bundles. *b* phloem. *c* cambium. *h* xylem. *P* pith. — (*Wiesner*.)

walls remain thin and are often furnished with simple pores. A good example of this type is furnished by different species of Salix. (Fig. 54.)

Euonymus and Berberis stems are examples of the other

class, where growth in thickness arises from the increase in size
of the original bundles, no new bundles being intercalated. This
kind of growth we have referred to as ring formation by interfascicular cambium. To describe this, we may assume a cross-
section so near the stem apex that the original bundles are yet
distinct and at some distance from each other. The continuous
ring of meristem cells is formed here by the cells lying on either
side of each bundle and in contact with the cambium, changing
or growing into cambium cells. Their number, reckoned radially,
corresponds to that of the cambium cells. This change in form
continues until only a few parenchymatic cells are left between
the bundles. These in turn become meristematic but retain
their parenchymatic form, and develop the primary medullary
rays, in this case the same in number as the original bundles.

A cross-section of a stem belonging to either of the above
types may be described as consisting of three parts, a central
circle of pith cells, followed by a zone of vascular tissue interrupted by the primary medullary rays, and succeeded by another
of ground tissue cells. All that growth which takes place previous to the completion of the cambium ring is called primary,
and the outer zone or that lying outside the cambium is
called the *primary rind*. When, on the completion and subsequent development of the ring, another entirely new zone is
formed, it is called *secondary* or *additional growth*. This
secondary zone always consists of two parts, an outer phloem,
and an inner wood part. The name secondary rind is sometimes applied to the phloem part of this secondary growth. In
pharmaceutical as well as in common language, the term rind
alone denotes all that part of the stem outside the cambium
ring.

The separate elements of the secondary growth, except the
medullary ray cells, have the same general character as those
corresponding in the primary growth. The cambium cells
themselves are long or prosenchymatic, with the long axis

parallel to that of the stem and somewhat sharpened ends, and lie in tangential rows of varying number. The middle row or rows are supposed to generate the new cells, but this is a very difficult matter to decide with certainty. The ducts are nearly all porous and the pores are bordered. Spiral ducts are very rare, but are found in certain Cacti forms. The primary bundles often develop these in such numbers as to form a sort of ring around the pith known as the pith crown. Tracheids are numerous, generally with small slit-formed pores like those of the libriform cells. The wood of Conifers is peculiar in that it produces no ducts in the secondary growth, their place being supplied by tracheids with considerably thickened walls and large bordered pores. The axis of the wood-parenchyma cell which is parallel with the long axis of the stem, is somewhat longer than the other two though not enough so to render it prosenchymatic. These cells generally occur in groups, and are situated not far from the ducts. Sometimes they surround them in a circle. Sometimes they form a nearly continuous circle around the stem. Together with the medullary ray cells they form the living elements of the wood. They originate from a prosenchymatic mother cell, which divides by anticlinal walls. The libriform cells have already been characterized in the description of the compound bundle.

The elements of the secondary phloem correspond to those of the primary, that is, they consist of bast fibers, sieve tubes and parenchyma. Besides these there are often groups of stone cells, and glands containing calcium oxalate crystals which often accompany the bast fibers. These last generally occur in groups. In some plants they are very long, in others shorter. The older layers of phloem generally have more bast than the younger, its formation often decreasing so that in later years only small bundles of it occur. Very few rinds are free from bast, while in many stems of trees the bast fibers build a closed cylindrical mantle.

The secondary parenchymatic and living element of the wood is the medullary ray cell which also forms a part of the phloem elements. The origin of the primary ray has been given and its function referred to. It is the only one extending from pith to rind and so connecting these portions. As each successive year adds a new zone of growth, the distance between these rays is constantly widening toward the circumference, and a new supply of living cells is added by the formation of new medullary rays. The cambium cells which are about to produce these divide by walls at right angles to their long axes, thus forming a parenchymatic meristem. The rays so formed are named *secondary*, or rays of the same order as the year in which they arise.

Regarding the character of the separate cells, their shape is similar to that of the wood parenchyma; their position, however,

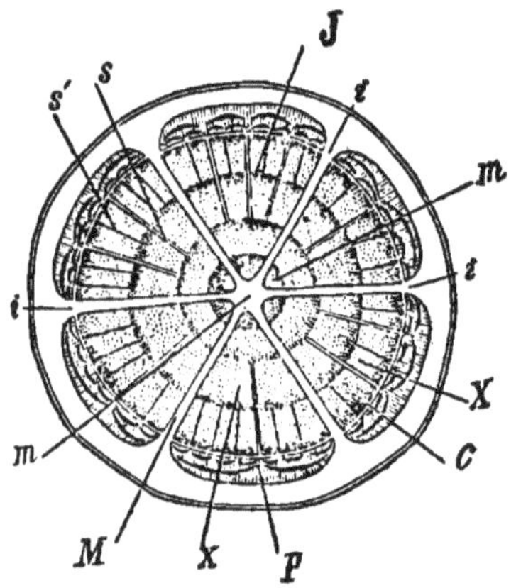

Fig. 53.

Diagrammatic cross-section through a three years' dicotyledonous stem (Type Berberis). *x* xylem. *p* phloem. *c* cambium. *i* interfasicular cambium. *m* pith. *m'* pith crown. *M* primary medullary rays. *s* secondary medullary rays of the first; *s'* of the second order. — (*Wiesner*.)

is reversed, so that the long axis is at right angles with that of the stem. Their walls are rather thick, unlignified, and richly supplied with simple pores. As a rule the primary rays have a greater width in the cross-section than the secondary; that is, they are often two, three, and even more cells wide, while the secondary are often only a single cell in width. The depth of the ray varies with the habit of the plant. Some are the depth of only a single cambium cell, others of several.

A cross-section of a stem of several years' growth presents the appearance of a circle composed of narrow, wedge-shaped sections, meeting at the centre. These are bounded by the medullary rays and contain the remaining elements of the wood. This xylem or wood portion also appears divided into as many separate zones as the stem has years of growth. These are commonly called *year's rings*, as, by a normal development, one is formed each year. The year's rings are a constant feature of all woody stems in which there is a cessation of activity during

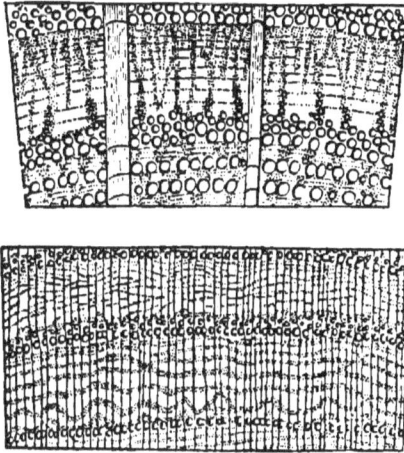

FIG. 56.

Cross-section of oak and elm wood showing arrangement of wood parenchyma and other elements.

the winter, the limits of each summer's growth being marked more or less distinctly by a change in the tissues produced. This change may be either in the kind of elements generated or in the different character of the same kinds. In the wood of most dicotyledons, the cells formed in the fall are thicker-walled with shortened radial diameter, while the spring wood is much richer in ducts, and the ducts have a greater diameter than those produced later in the season. In Conifers, which produce no secondary ducts, the years' rings are distinctly marked by the thickened walls and shortened radial diameter of the fall cells.

Trees growing in tropical climates, where there is little distinction of seasons and consequently no regular cessation of growth, lack this distinction of years' rings. (Fig. 56.)

While the cambium normally develops both phloem and xylem each year, the quantity of the latter is generally much greater than of the former. The phloem elements are also less regular in their arrangement and present little, if any, distinction between the different years' growths. It was on this account that the first explanation of the difference of the growth of spring and fall wood was, the pressure of the rind on the inner portion, which increased faster in radial diameter than the rind portion. This would account for the shortening of the radial diameter, but not for the increase of the thickness of wall, nor for the diminution in the number of ducts in the fall. Furthermore, it is now generally believed, from the results of numerous experiments, that the pressure of the rind in the fall is not, as a rule, greater than in the spring. While there are a number of different reasons offered to account for it, the reason for the difference between spring and fall growth is still a matter of discussion.

We have described the stem as constantly increasing in diameter by the intercalation of a hollow cylinder of new tissues; we must now inquire what is the effect of this on those portions which have been separated by this intercalary growth. It is evident at once that the central cylinder or pith must remain entirely unaffected by the new growth, or at most, suffer a little contraction from the pressure of the growing cylinder. It is different, however, with the condition of the outer cylinder or primary rind, which surrounds the growing tissues. It is easy to see that some means must be provided for its extension in the same ratio as that of the inner cylinder; otherwise it would be torn apart and destroyed by the increasing pressure from within. In stems with very limited secondary growth, the extension of the primary rind is often accomplished by a process of dilatation, by which the cells are increased in size without the formation of new cells. This extension is entirely insufficient to keep pace with the increasing growth of ordinary woody stems. For all such another plan is followed which results in what is called *periderm formation*.

The word periderm, is used to denote all that growth which is generated by a secondary meristem arising somewhere in the rind tissues. According to the position of this meristem, there are two kinds of periderm, superficial and deep-seated. On many stems of limited growth the superficial periderm is sufficient to supply the necessary extension, and also furnish a better means of protection than that given by the primary epidermis. In such stems no deep-seated periderm is formed. Where the woody growth is protected many years, the superficial periderm is deficient in the same manner as was the original epidermis, and therefore its place is supplied by the deep-seated.

There are three kinds of superficial periderm, the distinction here being based upon the place of the meristematic cells. These consist always of a single row of cells, as seen on the cross-section. In the superficial periderm they extend entirely around the stem thus forming a mantle or single-celled layer, unbroken save for the breathing spaces which will be described later. To this layer is given the name phellogen, and it arises either in the epidermal layer, in the second or third layer under this, or still deeper in the phloem of the vascular bundles. There are very few instances of the first class,[1] the second being the most common. As there is but a single layer of phellogen cells, their manner of division is more easily ascertained than in case of the cambium cells. In general, it may be either centripetal, centrifugal, or reciprocal; that is, the lasting cells may be cut off toward the centre, toward the circumference, or in both directions alternately. The cells cut off lie in regular radial rows; those external to the phellogen layer are somewhat plate-formed, their radial diameter remaining short. Their walls soon become

---

[1] Nerium Oleander may be used to illustrate.

suberized, and in this way corky layers are formed. Those within the phellogen layer, if any, grow to resemble the ground cells in form, and remain living, their walls never becoming suberized. Thus the outer or corky layers are suited to perform the functions of the primary epidermis, their suberized walls rendering them almost impervious to air and water; while the inner layers add to the radial diameter of the rind. The increase in circumference is supplied in both cases by the formation of radial walls and the subsequent growth of cells so formed in the phellogen layer.

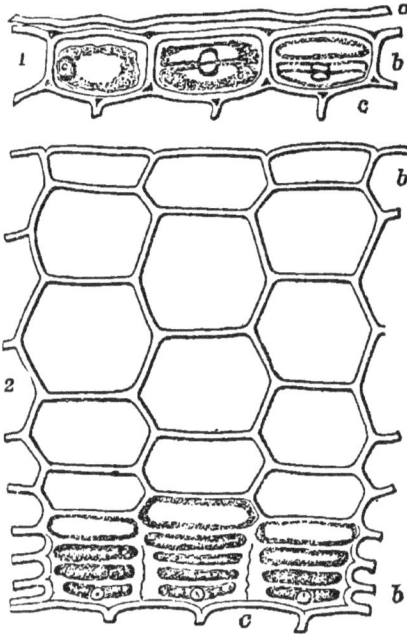

FIG. 57.

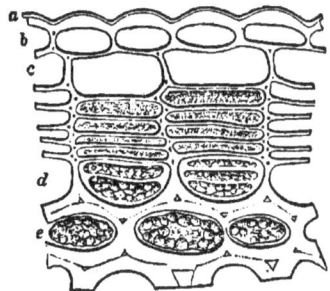

FIG. 58.

1. *a* the cuticle sprung off by the extension of the cellulose part of the wall below it. *b* the epidermal cells in process of division. *c* the outer wall of the outer rind layer.
2. The corky layers, *b b*, have been developed. The cells in the lower layer, *b*, which have a nucleus, are the phellogen cells. *c* the rind cells. — (*Th. Hartig.*)

a shows the cuticle; *b* epidermis with thickened outer wall; *c* the outermost cork cells which have originated from the layer of rind cells just under the epidermis; *d* the phellogen cells, and next to these the phelloderm containing chlorophyll. — (*Th. Hartig.*)

In all of the three classes described, there are suberized walls between the primary epidermis and the inner cells of the rind, upon which it depends for food. Lacking this it soon dies, and is cracked and worn off, leaving the outer periderm layers to take its place. In the second and third classes, that is also the fate of all the cells outside the corky layers. From this description it is evident that the farther from the circumference the phellogen layer originates, the better its adaptation for continued secondary growth. Take the first case for example; here the phellogen layer is as near the surface as possible, and all the rind layers except the epidermis are left intact between the inner layer of periderm cells and outer layer of the new cambium growth. The cells of these rind layers must increase their circumference or their tangential diameters to keep pace with the increasing circumference of both zones between which they lie, that is, the zones of the secondary growth from the cambium and the phellogen rings. This is probably accomplished, in part, by growth in surface of their periclinal walls, and in part, by the formation of new radial walls increasing the number of cells. In the second and third classes, the number of layers to which this growth is necessary regularly decreases.

Long continued growth from the cambium ring renders still other changes necessary, and this not only in the primary rind, but also in that part of the primary zone itself which lies nearest to the primary rind. This is supplied by the formation of the secondary or deep-seated periderm, which is called secondary because its formation is usually preceded by some form of superficial periderm.

It is much more difficult to give even a general description of the development of the deep-seated periderm, as it varies not only in different plants but on different organs of the same plant, and even on different parts of these organs. The reasons for such variety will be seen from the fact that the deep-seated or internal phellogen layer always arises deep enough in the rind to pass through or between all the elements of the phloem tissues, whether these tissues belong to the original bundles or to the subsequent growth of the cambium ring. The phellogen cells originate by certain living parenchymatic cells of this phloem tissue taking on the power to divide or form new walls. These living cells do not consist of a connected mass of tissue as is the case with the rind cells where the primary phellogen originates, but are separated by clusters of bast cells, stone cells, sieve-tubes, etc. For this reason the phellogen cells of the inner periderm do not appear to form a continuous ring, but vary in their course according to the variations in the order of arrangement of these clusters. Notwithstanding this apparent irregularity the result accomplished is similar; that is, the inner periderm separates from the inner portions of the stem a zone or hollow cylinder composed of various elements, instead of the simple ground tissue cut off by the superficial periderm. This zone or hollow cylinder comprises what is commonly called bark, namely all that portion of the stem outside the inner periderm.

This shows the difficulty of making a sharp distinction between the two classes of periderm, superficial and deep-seated. In fact the third kind of superficial periderm, where the phellogen originates in the phloem of the original bundles, sometimes serves the purposes of both, and in such cases no other is formed. A good example of this is the beech (Fagus). Here the first periderm is the only one, and by its constant development it keeps pace with the cylinder growing within.

In those plants where successive inner periderms are formed, if the first originates somewhat deep in the rind, the secondary periderms follow the course more or less closely. On the other hand, if the first originates near the circumference the secondary periderms follow successively in such a manner that taking a radially oblique course, they attach themselves to the older periderm, cutting off scale-shaped sections of bark between.

These periderms have been referred to as cutting off or separating outer portions of bark with no reference to an actual separation of the part cut off, except in the first instance where the primary epidermis is said to crack and become rubbed off.

There is, however, in many plants an actual separation of certain parts outside the periderm formation. (Fig. 59.) As before explained, all portions outside the periderm gradually die for lack of food. Not all, however, are removed from the plant, certain portions of bark remaining to perform the functions of epidermis, and some botanists refer to the bark as a third form of epidermis. In order to understand how some portions are

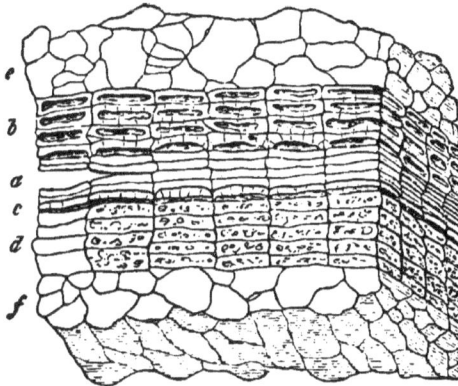

FIG. 59.

a cork layer separating bark scales. e and f large parenchymatic rind cells. b an outer zone of thick-walled, a one of thin-walled cork cells; the latter tears easily. c is a layer with only the outer wall thickened. d phelloderm.

separated and others are allowed to remain, it will be necessary to describe more fully the elements of the inner periderm which differ slightly from those already described. These elements may be tabulated as follows:

$$\text{Deep-seated Periderm.} \begin{cases} \text{Phellem} \begin{cases} \text{Cork.} \\ \text{Phelloid.} \end{cases} \\ \text{Phellogen} \quad \text{Initial or meristem layer.} \\ \text{Phelloderm} \begin{cases} \text{Inside layers, never corky} \\ \text{nor lignified.} \end{cases} \end{cases}$$

The term phellem here is used for all the tissue cut off toward the circumference; that part of it whose walls become lignified is called phelloid, while all the cells cut off centrip-

etally are named phelloderm ; these generally contain chlorophyll. The phellem cells, which form what is known as the
corky layer, are alike in form and origin ; they differ only in the
nature of their walls, some being entirely without suberin and
therefore either soft and easily torn asunder, or if their contents
are lost, brittle and easily broken. The separation or removal
of the sections of bark from the stem depends largely on this
lack of cohesion in the phelloid cells. Something also depends
on the nature of the tissues without. If these are tough and
of such a nature as to shrink forcibly with loss of water, their
contraction causes the tearing apart of the phelloid cells and
results in the loosening and separation of the bark sections which
they limit. The large scaly pieces of bark from Platanus stems,
and the long thin strips from Clematis and Vitis are separated
in this way, the former being called *scale bark*, the latter, *ring
bark*.

On the other hand, certain stems develop their periderm in
such a way that the breaking of the weaker walls takes place so
as to cause deep fissures or clefts in the bark without detaching
any large pieces. Examples of this are found in several species
of oak (Quercus). Still others form their periderm first only
on certain places, angled stems along the corners, as in certain
species of Euonymus, round stems in bands parallel with the
long axis, as in some species of Quercus. In all these instances
the phellogen produces only phellem and the masses so formed
are named wings. One peculiarity of this formation may have
a bearing on the question of the cause of difference between
spring and fall wood. The number of years' growth may be
detected in these cork ridges quite as readily as in the woody
portion, the regular summer's growth being terminated always
by several layers of cells with exceedingly short radial diameter.
This wing-like growth is mostly confined to the younger stems
and branches. Later on the entire surface of the stem is
covered by the gradual extension of the phellogen layer and

the periderm is developed in the ordinary manner. The cork wings are gradually worn away. Stems of Liquidamber illustrate this.

The outer walls of the primary epidermal cells taken collectively form a membrane around the plant organ which is more

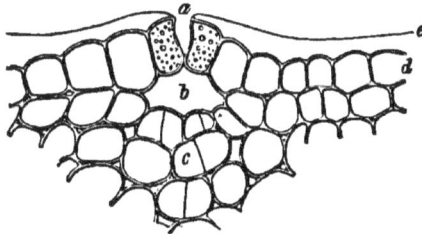

FIG. 60.

Beginning of the growth of a lenticel on a birch branch. *a* stoma. *b* air space. *c* beginning of divisions which give rise to the filling tissue. *d* epidermal cells. *e* cuticle raised up. — (*De Bary.*)

or less impervious to air and water. The corky layers of the secondary epidermis or periderm form a covering with similar

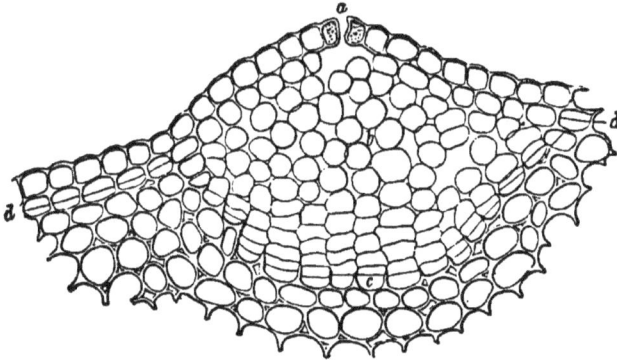

FIG. 61.

Finished lenticel. *a* stoma. *b* intercellular space in the filling tissue. *e* lenticel-phellogen. *d* beginning of periderm. — (*De Bary.*)

function. This latter covering is supplied with breathing places or pores corresponding to the stomata of the primary epidermis; to these organs De Candolle gave the name lenticel.

The lenticels of the superficial periderm arise generally in the ground tissue under a stoma or a group of the same. They are usually lens-shaped or oval with a fissure running lengthwise through the middle. They are generally raised above the surface of the periderm, more rarely sunk below or on a level with it. Many are extremely small, appearing only as small points. The greater number are larger, reaching sometimes more than a centimeter in length. Examples, lenticels of Betula.

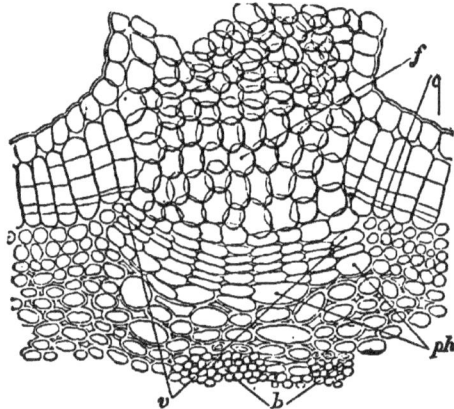

FIG. 62.

Lenticel from *Sambucus nigra.* *f* filling cells. *c* phellogen. *v* rejuvenating layer. *ph* phelloderm. *b* bast bundle. — (*Accd. to Stahl.*)

In some instances lenticels originate before the formation of periderm, by certain of the ground cells below a stoma becoming meristematic and forming a phellogen layer which eventually joins that of the periderm. (Figs. 60 and 61). This phellogen develops parenchymatic tissue in both directions, that toward the center becoming ordinary chlorophyll-holding cells, while that without forms what is known as the filling tissue. This consists at first of ordinary isodiametric cells, which after a time separate at the corners and finally become quite free. By their rapidly increasing numbers (Fig. 62) they crowd up into the air space, and finally rupture the epidermis, thus leaving a free communication between the air outside and the chloro-phyll-holding cells within. At the close of the summer, when the time for the winter rest draws near, this opening is usually closed by the last layers of the phellogen growth. That is, these layers remain a continuous collection of cells and their walls become suberized. In the spring time, the phellogen renews its activity, and the new cells, pushing up as before, break these layers apart and the opening is established anew. (Fig. 63.)

On stems and roots where the first periderm is replaced by bark, the communication necessary for breathing is sometimes

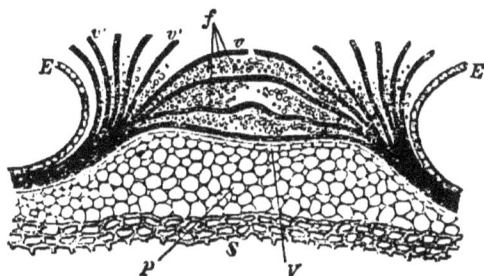

FIG. 63.

Lenticel from a branch of Gleditschia, with several closing layers, *r*. *E* epidermis. *r'* closing layers broken off. *l'* rejuvenating layer. *f* filling cells. *p* parenchyma. *s* schlerenchyma.

made by the fissures and rifts of the bark extending through the corky covering; and in other cases by the growth of lenticels originating with no reference to the place of the original stomata. These lenticels originate, in some instances, under the fissures, so they open into them; in others they occur under the ridges and extend through the entire thickness of the bark and open on the ridges. The central portion of woody tissue in stems of considerable age is known as *heart wood*. Its elements are dead and its color is usually darker than that of the surrounding cyl-inder, which is called *splint wood*. This latter is that portion of the wood which conducts the water currents upward, and in which the parenchymatic and medullary ray cells are still living.

ABNORMAL CASES OF SECONDARY GROWTH.

Secondary growth occurs so seldom among the monocotyledons that the few cases of it may be classed with the abnormal cases of the dicotyledons. Among the latter class there are many variations from the type described; a few prominent ones may be mentioned here. In certain genera, the cambium of the ring loses power to grow after a time and another cambium ring originates outside it. In Cocculus laurifolius and some others of the Menispermaceae this new ring forms in the primary rind; in Phytolacca, Wisteria and some species of Bauhinea, in the phloem tissue of the primary bundles. This new ring develops phloem outside and xylem inside just as the old one did, and may in its turn be replaced by a succession of other rings.

Some of the Cycads form outer rings in a similar manner, making a complicated structure. *Tecoma radicans* develops a cambium ring on the inside next the pith, which cuts off new cells in contrary directions from those of the first or normal cambium ring, that is, phloem inward and xylem outward. In *Mirabilis Jalapa* the leaf-trace strands form a somewhat irregular collection deep in the pith, while the cambium ring is formed from stem bundles.

Rhizomes of certain species of Rheum have a singular structure, the leaf-trace bundles forming a normal ring which develops regularly, enclosing a relatively large pith. Through this pith, in the early stage, run other vascular bundles obliquely, anastomosing with each other and with the leaf-trace bundles. At first these consist simply of sieve-tubes and their accompanying cells; later they surround themselves each with a cambium ring which develops phloem inside and xylem outside.

Still more remarkable forms are seen in the climbing stems of Sapindaceae. Sarjania is the one best known. Here on a cross-section, there are one central and from five to seven peripherical wood rings, each having its own axis. In the centre of each woody growth lies a large-celled pith. In the wood are large porous ducts, and outside the cambium a narrow bast rind. This latter becomes contorted through the pressure exerted by the growing wood portions. Of the monocotyledons, only a few have secondary growth. Some of the Liliaceae which have trunk-like stems, Dracaena, Cardyline, Yucca, and Aloe form in the periphery of their central cylinder a cambium ring from which are cut off toward the centre isolated xylem bundles with parenchyma between them. Toward the outer surface phloem is developed and later a bark-like covering.

FIGURE SHOWING NORMAL SECONDARY GROWTH OF CONIFER WOOD.

FIG. 64.

Figure 64 represents solid sections of Conifer wood, show-ing its normal manner of growth. Number 1 is a triangular section from pith to rind showing ten years' rings. The upper portion shows the rings, while the face view gives the appear-ance of the medullary rays as seen on a long radial section. Number 2 is an enlarged portion showing the anatomical characters of a single year's growth. The medullary rays are here represented as being only a single cell in width, as seen on the cross-section and also on the tangential long section represented by the right-hand face view.

INDEX.

Introduction to Physical Science.

By A. P. GAGE, Ph.D., Instructor in Physics in the English High School, Boston, Mass. 12mo. Cloth. viii + 353 pages. With a color chart of spectra, etc. Mailing price, $1.10; for introduction, $1.00.

THE long-continued increasing popularity of Gage's *Elements of Physics* has created a demand for an easier book, on the same plan, suited to schools that can give but a limited time to the study. The *Introduction to Physical Science* meets this demand.

In a text-book, the first essentials are correctness and accuracy. It is believed that Gage's *Introduction* will stand the closest expert scrutiny. Especial care has been taken to restrict the use of scientific terms, such as *force, energy, power*, etc., to their proper significations. Recent advances in physics have been faithfully recorded. Among the new features are a full treatment of electric lighting, and descriptions of storage batteries, methods of transmitting electric energy, simple and easy methods of making electrical measurements with inexpensive apparatus, the compound steam-engine, etc. Static electricity, now generally regarded as of comparatively little practical importance, is treated briefly; while dynamic electricity, the most promising physical agent of modern times, is placed in the clearest light of our present knowledge.

The style will be found suited to the grades that will use the book. The experiments are of **practical** significance, and simple in manipulation.

The *Introduction*, like the author's *Elements*, has this distinct and distinctive aim, — to elucidate science, instead of "popularizing" it; to make it liked for its own sake, rather than for its gilding and coating; and, while teaching the facts, to impart the spirit of science, that is to say, the **spirit** of our civilization and progress.

J. P. Naylor, *Professor of Physics, De Pauw University:* In its scientific spirit, and in accuracy and clearness of statements of principles, I know of nothing that is its superior.

O. C. Kinyon, *Teacher of Physics in High School, Syracuse, N. Y.:* It not only insures an interest in the study, but tends to thoroughly arouse the powers of observation.

B. C. Hinde, *Professor of Natural Science, Trinity College, N.C.:* I have used Gage's. It is strictly in accord with the best modern teaching of Physics.

Introduction to Chemical Science.

By R. P. WILLIAMS, Instructor in Chemistry in the English High School, Boston. 12mo. Cloth. 216 pages. By mail, 90 cents; for introduction, 80 cents.

THIS work is strictly, but easily, inductive. The pupil is stimulated by query and suggestion to observe important phenomena, and to draw correct conclusions. The experiments are illustrative, the apparatus is simple and easily made. The nomenclature, symbols, and writing of equations are made prominent features. In descriptive and theoretical chemistry, the arrangement of subjects is believed to be especially superior in that it presents, not a mere aggregation of facts, but the *science* of chemistry. Brevity and concentration, induction, clearness, accuracy, and a legitimate regard for interest, are leading characteristics. The treatment is full enough for any high school or academy.

Though the method is an advanced one, it has been so simplified that pupils experience no difficulty, but rather an added interest, in following it.

The author himself has successfully employed this method in classes so large that the simplest and most practical plan has been a necessity.

Thomas C. Van Nuys, *Professor of Chemistry, Indiana University, Bloomington, Ind.:* I consider it an excellent work for students entering upon the study of chemistry.

C. F. Adams, *Teacher of Science, High School, Detroit, Mich.:* I have carried two classes through Williams's Chemistry. The book has surpassed my highest expectations. It gives greater satisfaction with each succeeding class.

J. W. Simmons, *County Superintendent of Schools, Owosso, Mich.:* The proof of the merits of a textbook, is found in the crucible of the class-room work. There are many chemistries, and good ones; but, for our use, this leads them all. It is stated in language plain, interesting and not misleading. A logical order is followed, and the mind of the student is at work because of the many suggestions offered. We use Williams's work, and the results are all we could wish. There is plenty of chemistry in the work for any of our high schools.

W. J. Martin, *Professor of Chemistry, Davidson College, N.C.:* One of the most admirable little textbooks I have ever seen.

T. H. Norton, *Professor of Chemistry, Cincinnati University, O.:* Its clearness, accuracy, and compact form render it exceptionally well adapted for use in high and preparatory schools. I shall warmly recommend it for use, whenever the effort is made to provide satisfactory training in accordance with the requirements for admission to the scientific courses of the University.

Scheiner's Astronomical Spectroscopy.

Department of Special Publication. — Translated, revised and enlarged by E. B. FROST, Associate Professor of Astronomy in Dartmouth College. 8vo. Half leather. Illustrated. xiii + 482 +   pages. Price by mail, $5.00; for introduction, $4.75.

THIS work aims to explain the most practical and modern methods of research, and to state our present knowledge of the constitution, physical condition and motions of the heavenly bodies, as revealed by the spectroscope.

There are three parts : — I. Spectroscopic Apparatus ; II. Spectral Theories ; III. Results of Spectroscopic Observations, with a fourth containing tables of wave-lengths of lines of the solar spectrum, catalogues of stars with special types of spectra, and a full bibliography brought down to 1893.

· **Edward S. Holden**, *Director of the Lick Observatory, Mt. Hamilton, California:* I congratulate you on the appearance of this very important book; it is indispensable to all astronomers and students of spectroscopy.

W. W. Campbell, *Astronomer, Lick Observatory, University of California:* I have only words of praise to bestow upon the book.

Elements of Structural and Systematic Botany.

For High Schools and Elementary College Courses. By DOUGLAS H. CAMPBELL, Professor of Botany in the Leland Stanford Junior University. 12mo. Cloth. ix + 253 pages. Price by mail, $1.25; for introduction, $1.12.

THE fundamental peculiarity and merit of this book is that it begins with the simple forms, and follows the order of nature to the complex ones.

R. Ellsworth Call, *Teacher of Natural Science, High School, West Des Moines, Ia.:* It is the only manual which combines just enough technic with a logical treatment of the more minute structure of plants, to render it a help to the teacher of to-day.

Plant Organization.

By R. HALSTED WARD, M.D., F.R.M.S., Professor of Botany in the Rensselaer Polytechnic Institute, Troy, N.Y. Quarto. 176 pages. Illustrated. Flexible boards. Mailing price, 85 cents; for introduction, 75 cents.

IT consists of a synoptical review of the general structure and morphology of plants, with blanks for written exercises by pupils.

NATURAL SCIENCE TEXT-BOOKS.

ELEMENTS OF PHYSICS. A Text-book for High Schools and Academies. By ALFRED P. GAGE, A.M., Instructor in Physics in the English High School, Boston. $1.12.

C. F. Emerson, *Prof. of Physics, Dartmouth College :* " It takes up the subject on the right plan, and presents it in a clear yet scientific way."

INTRODUCTION TO PHYSICAL SCIENCE. By A. P. GAGE, author of " Elements of Physics." $1.00.

B. F. Sharpe, *Prof. of Natural Science, Randolph-Macon College, Va. :* " It is the very thing for the academy preparatory to this college."

PHYSICAL LABORATORY MANUAL AND NOTE-BOOK. By A. P. GAGE, author of " Elements of Physics," " Introduction to Physical Science," etc. 35 cents.

I. Thornton Osmond, *Prof. of Physics, Penn. State College :* " It is a product of the ability, experience, and sound judgment that have made Dr. Gage's other books the best of their rank in physics."

INTRODUCTION TO CHEMICAL SCIENCE. By R. P. WILLIAMS, Instructor in Chemistry in the English High School, Boston. 80 cents.

Arthur B. Willmot, *Prof. of Chemistry, Antioch College, Ohio :* " It is the best chemistry I know of for high-school work."

LABORATORY MANUAL OF GENERAL CHEMISTRY. By R. P. WILLIAMS, author of " Introduction to Chemical Science." 25 cents.

W. M. Stine, *Prof. of Chemistry, Ohio University, Athens, Ohio :* " It is a work that has my heartiest indorsement. I consider it thoroughly pedagogical in its principles."

YOUNG'S GENERAL ASTRONOMY. A Text-book for Colleges and Technical Schools. By CHARLES A. YOUNG, Ph.D., LL.D., Prof. of Astronomy in Princeton College, and author of " The Sun," etc. $2.25.

S. P. Langley, *Sec. Smithsonian Institution, Wash., D.C.,* and *Pres. National Academy of Sciences :* " I know no better book (not to say as good a one) for its purpose on the subject."

YOUNG'S ELEMENTS OF ASTRONOMY. A Text-book for Use in High Schools and Academies, with a Uranography. By CHARLES A. YOUNG, author of " Young's General Astronomy," " The Sun," etc. $1.40. **Uranography.** From " Young's Elements of Astronomy." 30 cents.

S. H. Brickett, *Teacher of Mathematics, St. Johnsbury Academy, Vt. :* " It is just what I expected it would be, the very best which I have ever seen."

YOUNG'S LESSONS IN ASTRONOMY. Including Uranography. By CHARLES A. YOUNG, author of " A General Astronomy," " Elements of Astronomy," etc. Prepared for schools that desire a brief course free from mathematics. $1.20.

AN INTRODUCTION TO SPHERICAL AND PRACTICAL ASTRONOMY. By DASCOM GREENE, Prof. of Mathematics and Astronomy in the Rensselaer Polytechnic Institute, Troy, N.Y. $1.50.

Davis Garber, *Prof. of Astronomy, Muhlenberg College :* " Students pursuing astronomy on a practical line will find it a very excellent and useful book."

ELEMENTS OF STRUCTURAL AND SYSTEMATIC BOTANY. For High Schools and Elementary College Courses. By DOUGLAS HOUGHTON CAMPBELL, Ph.D., Prof. of Botany in the Indiana University. $1.12.

Charles W. Dodge, *Teacher of Botany, High School, Detroit, Mich. :* " It is the only English work at all satisfactory for high-school students."

BLAISDELL'S PHYSIOLOGIES: Our Bodies and How We Live, 65 cents ; **How to Keep Well,** 45 cents ; **Child's Book of Health,** 30 cents.

True, scientific, interesting, teachable.

ELEMENTARY METEOROLOGY. By WILLIAM M. DAVIS, Prof. of Physical Geography in Harvard University. With maps, charts, and exercises. $2.50.

Copies will be sent, post paid, to teachers for examination on receipt of the introduction prices given above.

GINN & COMPANY, Publishers.

BOSTON. NEW YORK. CHICAGO. LONDON.

www.ingramcontent.com/pod-product-compliance
Lightning Source LLC
Chambersburg PA
CBHW021811190326
41518CB00007B/539